MW00608890

FIVE HUNDRED AND SEVEN

MECHANICAL MOVEMENTS,

EMBRACING

ALL, THOSE WHICH ARE MOST IMPORTANT IN

DYNAMICS, HYDRAULICS, HYDROSTATICS, PNEUMATICS, STEAM
ENGINES, MILL AND OTHER GEARING, PRESSES, HOROLOGY,
AND MISCELLANEOUS MACHINERY;

AND INCLUDING

MANY MOVEMENTS NEVER BEFORE PUBLISHED,

AND

SEVERAL WHICH HAVE ONLY RECENTLY COME INTO USE.

BY

HENRY T. BROWN,

EDITOR OF THE "AMERICAN ARTISAN."

———◆———

NEW-YORK:

PUBLISHED BY BROWN, COOMBS & CO.,

OFFICE OF THE "AMERICAN ARTISAN,"

189 BROADWAY.

1871.

PREFACE.

THE want of a comprehensive collection of illustrations and descriptions of ME-CHANICAL MOVEMENTS has long been seriously felt by artisans, inventors, and students of the mechanic arts. It was the knowledge of this want which induced the compilation of the collection here presented. The movements which it contains have been already illustrated and described in occasional installments scattered through five volumes of the AMERICAN ARTISAN, by the readers of which their publication was received with so much favor as was believed to warrant the expense of their reproduction with some revision in a separate volume.

The selection of the movements embraced in this collection has been made from many and various sources. The English works of Johnson, Willcock, Wylson, and Denison have been drawn upon to a considerable extent, and many other works—American and foreign—have been laid under contribution; but more than one-fourth of the movements—many of purely American origin—have never previously appeared in any published collection. Although the collection embraces about three times as many movements as have ever been contained in any previous American publication, and a considerably larger number than has ever been contained in any foreign one, it has not been the object of the compiler to merely swell the number, but he has endeavored to select only such as may be of really practical value; and with this end in view, he has rejected many which are found in nearly all the previously published collections, but which he has considered only applicable to some exceptional want.

Owing to the selection of these movements at such intervals as could be snatched from professional duties, which admitted of no postponement, and to the engravings having been made from time to time for immediate publication, the classification of the movements is not as perfect as the compiler could have desired; yet it is believed that this deficiency is more than compensated for by the copiousness of the *Index* and the entirely novel arrangement of the illustrations and the descriptive letter-press on opposite pages, which make the collection—large and comprehensive as it is—more convenient for reference than any previous one.

INDEX

- - •••

☞ IN this INDEX the numerals do not indicate the pages, but they refer to *engravings* and the *numbered paragraphs*. Each page of the letter-press contains the descriptive matter appertaining to the illustrations which face it.

———— •••• ————

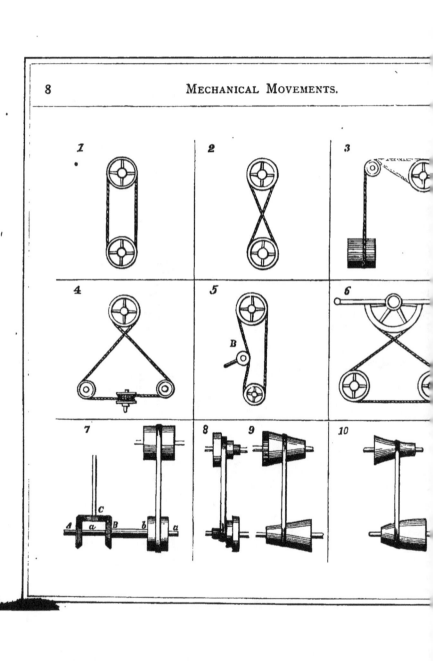

1. Illustrates the transmission of power by simple pulleys and an open belt. In this case both of the pulleys rotate in the same direction.

2. Differs from 1 in the substitution of a crossed belt for the open one. In this case the direction of rotation of the pulleys is reversed.

By arranging three pulleys, side by side, upon the shaft to be driven, the middle one fast and the other two loose upon it, and using both an open and a crossed belt, the direction of the said shaft is enabled to be reversed without stopping or reversing the driver. One belt will always run on the fast pulley, and the other on one of the loose pulleys. The shaft will be driven in one direction or the other, according as the open or crossed belt is on the fast pulley.

3. A method of transmitting motion from a shaft at right angles to another, by means of guide-pulleys. There are two of these pulleys, side by side, one for each leaf of the belt.

4. A method of transmitting motion from a shaft at right angles to another whose axis is in the same plane. This is shown with a crossed belt. An open belt may be used, but the crossed one is preferable, as it gives more surface of contact.

5. Resembles 1, with the addition of a movable tightening pulley, B. When this pulley is pressed against the band to take up the slack, the belt transmits motion from one of the larger pulleys to the other; but when it is not, the belt is so slack as not to transmit motion.

6. By giving a vibratory motion to the lever secured to the semi-circular segment, the belt attached to the said segment imparts a reciprocating rotary motion to the two pulleys below.

7. A method of engaging, disengaging, and reversing the upright shaft at the left. The belt is shown on the middle one of the three pulleys on the lower shafts, a, b, which pulley is loose, and consequently no movement is communicated to the said shafts. When the belt is traversed on the left-hand pulley, which is fast on the hollow shaft, b, carrying the bevel-gear, B, motion is communicated in one direction to the upright shaft; and on its being traversed on to the right-hand pulley, motion is transmitted through the gear, A, fast on the shaft, a, which runs inside of b, and the direction of the upright shaft is reversed.

8. Speed-pulleys used for lathes and other mechanical tools, for varying the speed according to the work operated upon.

9. Cone-pulleys for the same purpose as 8. This motion is used in cotton machinery, and in all machines which are required to run with a gradually increased or diminished speed.

10. Is a modification of 9, the pulleys being of different shape.

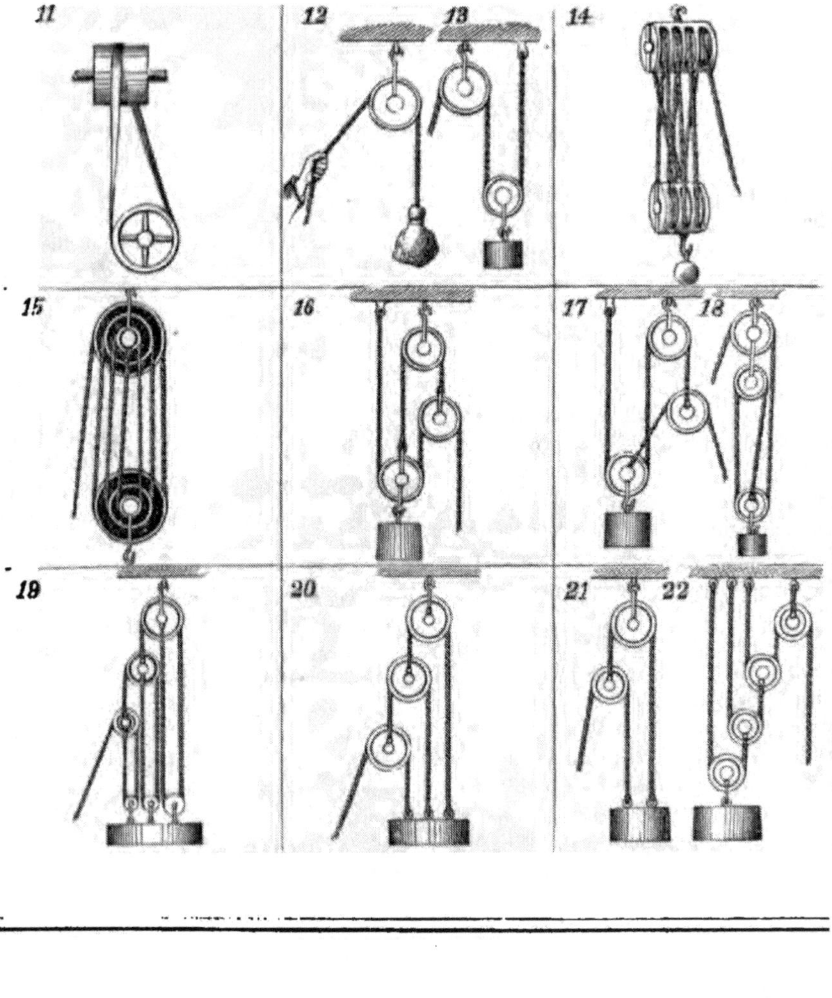
11
12
13
14
15
16
17
18
19
20
21
22

11. Another method of effecting the same result as 3, without guide-pulleys.

12. Simple pulley used for lifting weights. In this the power must be equal to the weight to obtain equilibrium.

13. In this the lower pulley is movable. One end of the rope being fixed, the other must move twice as fast as the weight, and a corresponding gain of power is consequently effected.

14. Blocks and tackle. The power obtained by this contrivance is calculated as follows: Divide the weight by double the number of pulleys in the lower block; the quotient is the power required to balance the weight.

15. Represents what are known as White's pulleys, which can either be made with sep- arate loose pulleys, or a series of grooves can be cut in a solid block, the diameters being made in proportion to the speed of the rope; that is, 1, 3, and 5 for one block, and 2, 4, and 6 for the other. Power as 1 to 7.

16 and 17. Are what are known as Spanish bartons.

18. Is a combination of two fixed pulleys and one movable pulley.

19, 20, 21, and 22. Are different arrangements of pulleys. The following rule applies to these pulleys :—In a system of pulleys where each pulley is embraced by a cord attached at one end to a fixed point and at the other to the center of the movable pulley, the effect of the whole will be = the number 2, multiplied by itself as many times as there are movable pulleys in the system.

23

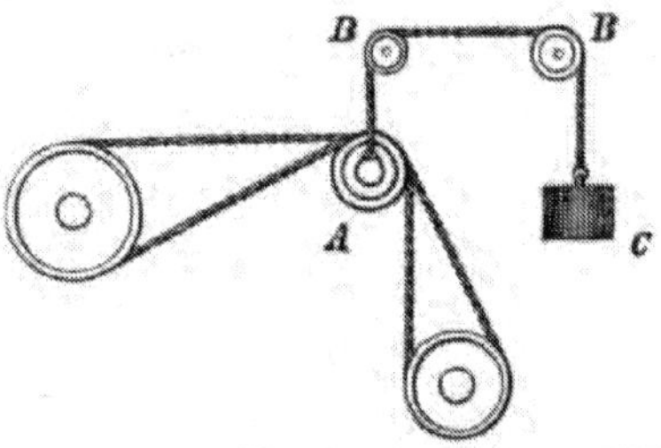

24

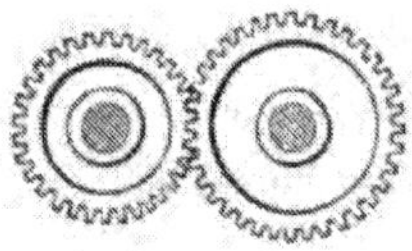

25

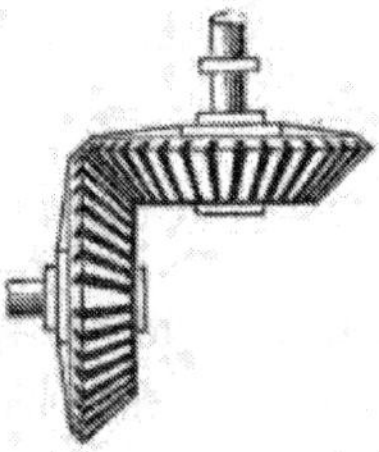

26

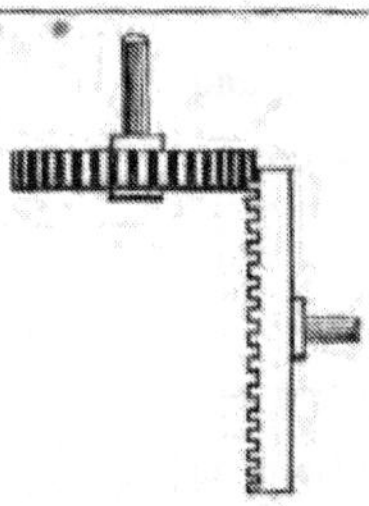

27

28

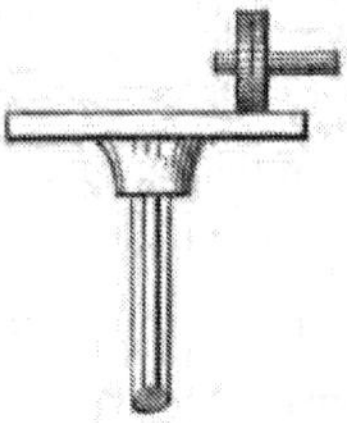

29

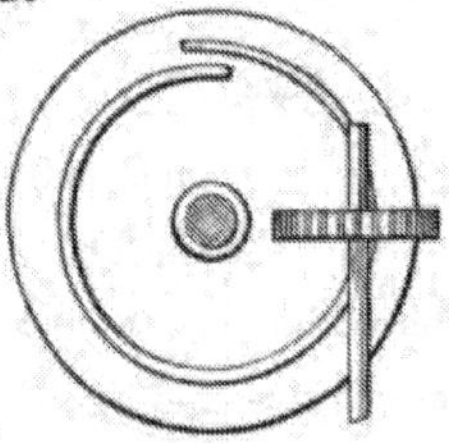

30

23. A contrivance for transmitting rotary motion to a movable pulley. The pulley at the bottom of the figure is the movable one; if this pulley were raised or depressed, the belt would be slackened or tightened accordingly. In order to keep a uniform tension on the belt, a pulley, A, carried in a frame sliding between guides (not shown), hangs from a rope passing over the two guide-pulleys, B, B, and is acted upon by the balance weight, C, in such manner as to produce the desired result.

24. Spur-gears.

25. Bevel-gears. Those of equal diameters are termed "miter-gears."

26. The wheel to the right is termed a "crown-wheel;" that gearing with it is a spur-gear. These wheels are not much used, and are only available for light work, as the teeth of the crown-wheel must necessarily be thin.

27. "Multiple gearing"—a recent invention. The smaller triangular wheel drives the larger one by the movement of its attached friction-rollers in the radial grooves.

28. These are sometimes called "brush-wheels." The relative speeds can be varied by changing the distance of the upper wheel from the center of the lower one. The one drives the other by the friction or adhesion, and this may be increased by facing the lower one with india-rubber.

29. Transmission of rotary motion from one shaft at right angles to another. The spiral thread of the disk-wheel drives the spur-gear, moving it the distance of one tooth at every revolution.

30. Rectangular gears. These produce a rotary motion of the driven gear at a varying speed. They were used on a printing-press, the type of which were placed on a rectangular roller.

31

32

33

34

35

36

37

38

39

31. Worm or endless screw and a worm-wheel. This effects the same result as 29; and as it is more easily constructed, it is oftener used.

32. Friction-wheels. The surfaces of these wheels are made rough, so as to *bite* as much as possible ; one is sometimes faced with leather, or, better, with vulcanized india-rubber.

33. Elliptical spur-gears. These are used where a rotary motion of varying speed is required, and the variation of speed is determined by the relation between the lengths of the major and minor axes of the ellipses.

34. An internally toothed spur-gear and pinion. With ordinary spur-gears (such as represented in 24) the direction of rotation is opposite ; but with the internally toothed gear, the two rotate in the same direction ; and with the same strength of tooth the gears are capable of transmitting greater force, because more teeth are engaged.

35. Variable rotary motion produced by uniform rotary motion. The small spur-pinion works in a slot cut in the bar, which turns loosely upon the shaft of the elliptical gear. The bearing of the pinion-shaft has applied to it a spring, which keeps it engaged ; the slot in the bar is to allow for the variation of length of radius of the elliptical gear.

36. Mangle-wheel and pinion—so called from their application to mangles—converts continuous rotary motion of pinion into reciprocating rotary motion of wheel. The shaft of pinion has a vibratory motion, and works in a straight slot cut in the upright stationary bar to allow the pinion to rise and fall and work inside and outside of the gearing of the wheel. The slot cut in the face of the mangle-wheel and following its outline is to receive and guide the pinion-shaft and keep the pinion in gear.

37. Uniform into variable rotary motion. The bevel-wheel or pinion to the left has teeth cut through the whole width of its face. Its teeth work with a spirally arranged series of studs on a conical wheel.

38. A means of converting rotary motion, by which the speed is made uniform during a part, and varied during another part, of the revolution.

39. Sun-and-planet motion. The spur-gear to the right, called the planet-gear, is tied to the center of the other, or sun-gear, by an arm which preserves a constant distance between their centers. This was used as a substitute for the crank in a steam engine by James Watt, after the use of the crank had been patented by another party. Each revolution of the planet-gear, which is rigidly attached to the connecting-rod, gives two to the sun-gear, which is keyed to the fly-wheel shaft.

40
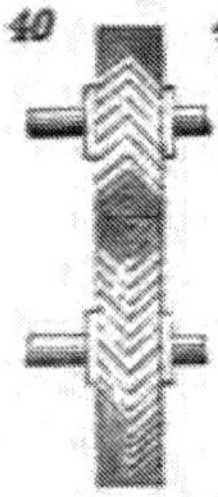
41
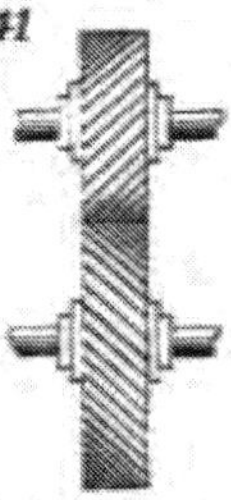
42
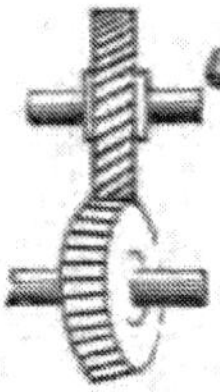
43
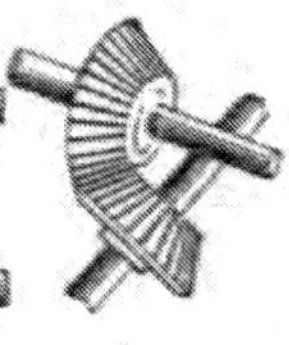
44
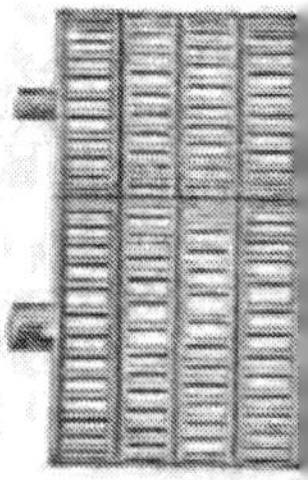
45
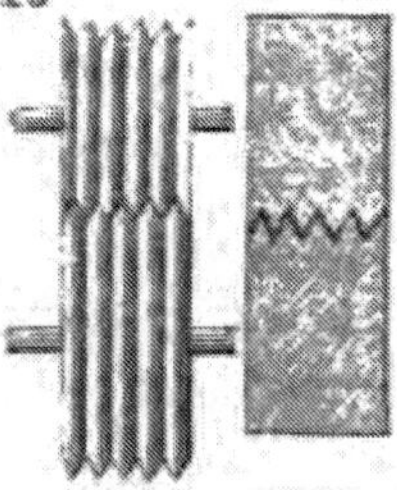
46
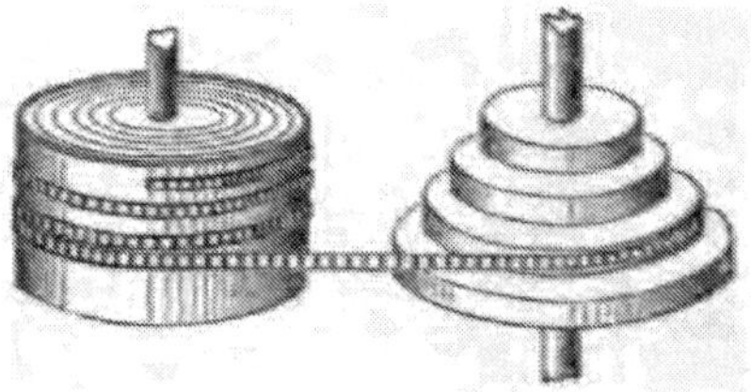
47
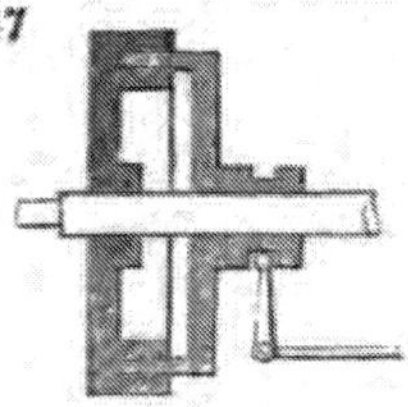
48
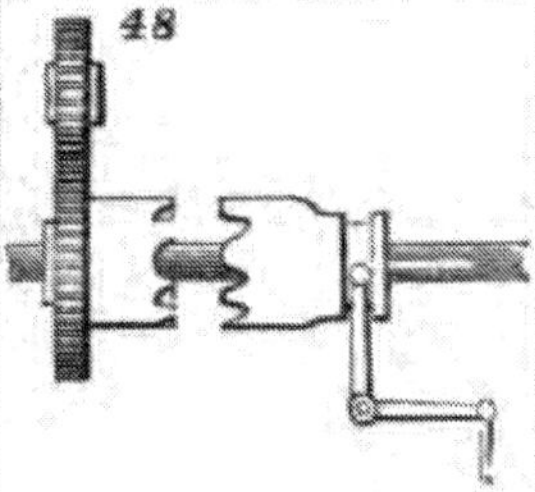
49
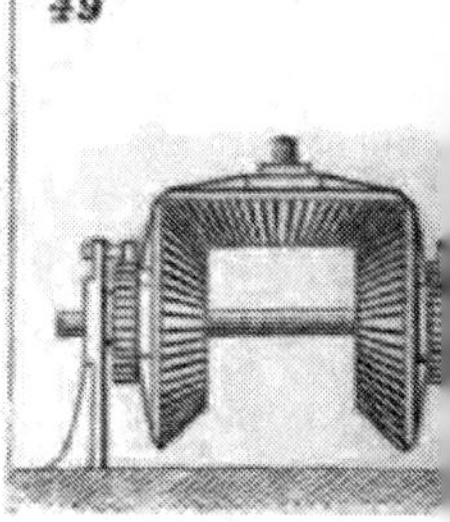

40 and 41. Rotary converted into rotary motion. The teeth of these gears, being oblique, give a more continuous bearing than ordinary spur-gears.

42 and 43. Different kinds of gears for transmitting rotary motion from one shaft to another arranged obliquely thereto.

44. A kind of gearing used to transmit great force and give a continuous bearing to the teeth. Each wheel is composed of two, three, or more distinct spur-gears. The teeth, instead of being in line, are arranged in steps to give a continuous bearing. This system is sometimes used for driving screw propellers, and sometimes, with a rack of similar character, to drive the beds of large iron-planing machines.

45. Frictional grooved gearing—a comparatively recent invention. The diagram to the right is an enlarged section, which can be more easily understood.

46. Fusee chain and spring-box, being the prime mover in some watches, particularly of English make. The fusee to the right is to compensate for the loss of force of the spring as it uncoils itself. The chain is on the small diameter of the fusee when the watch is wound up, as the spring has then the greatest force.

47. A frictional clutch-box, thrown in and out of gear by the lever at the bottom. This is used for connecting and disconnecting heavy machinery. The eye of the disk to the right has a slot which slides upon a long key or feather fixed on the shaft.

48. Clutch-box. The pinion at the top gives a continuous rotary motion to the gear below, to which is attached half the clutch, and both turn loosely on the shaft. When it is desired to give motion to the shaft, the other part of the clutch, which slides upon a key or feather fixed in the shaft, is thrust into gear by the lever.

49. Alternate circular motion of the horizontal shaft produces a continuous rotary motion of the vertical shaft, by means of the ratchet-wheels secured to the bevel-gears, the ratchet-teeth of the two wheels being set opposite ways, and the pawls acting in opposite directions. The bevel-gears and ratchet-wheels are loose on the shaft, and the pawls attached to arms firmly secured on the shaft.

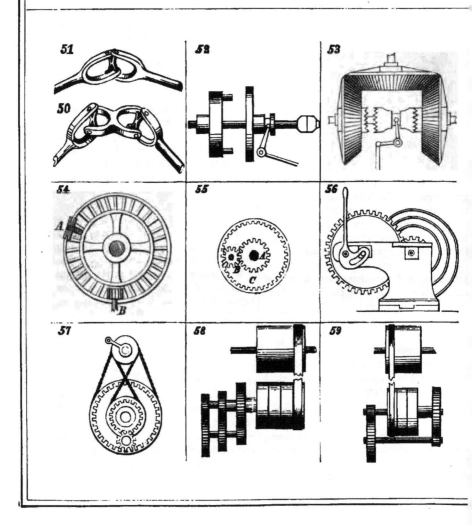

50 and 51. Two kinds of universal joints.

52. Another kind of clutch-box. The disk-wheel to the right has two holes, corresponding to the studs fixed in the other disk; and, being pressed against it, the studs enter the holes, when the two disks rotate together.

53. The vertical shaft is made to drive the horizontal one in either direction, as may be desired, by means of the double-clutch and bevel-gears. The gears on the horizontal shaft are loose, and are driven in opposite directions by the third gear; the double-clutch slides upon a key or feather fixed on the horizontal shaft, which is made to rotate either to the right or left, according to the side on which it is engaged.

54. Mangle or star-wheel, for producing an alternating rotary motion.

55. Different velocity given to two gears, A and C, on the same shaft, by the pinion, B.

56. Used for throwing in and out of gear the speed-motion on lathes. On depressing the lever, the shaft of the large wheel is drawn backward by reason of the slot in which it slides being cut eccentrically to the center or fulcrum of the lever,

57. The small pulley at the top being the driver, the large, internally-toothed gear and the concentric gear within will be driven in opposite directions by the bands, and at the same time will impart motion to the intermediate pinion at the bottom, both around its own center and also around the common center of the two concentric gears.

58. For transmitting three different speeds by gearing. The lower part of the band is shown on a loose pulley. The next pulley is fixed on the main shaft, on the other end of which is fixed a small spur-gear. The next pulley is fixed on a hollow shaft running on the main shaft, and there is secured to it a second spur-gear, larger than the first. The fourth and last pulley to the left is fixed on another hollow shaft running loosely on the last-named, on the other end of which is fixed the still larger spur-gear nearest to the pulley. As the band is made to traverse from one pulley to another, it transmits three different velocities to the shaft below.

59. For transmitting two speeds by gearing. The band is shown on the loose pulley—the left-hand one of the lower three. The middle pulley is fixed on the same shaft as the small pinion, and the pulley to the right on a hollow shaft, on the end of which is fixed the large spur-gear. When the band is on the middle pulley a slow motion is transmitted to the shaft below; but when it is on the right-hand pulley a quick speed is given, proportionate to the diameter of the gears.

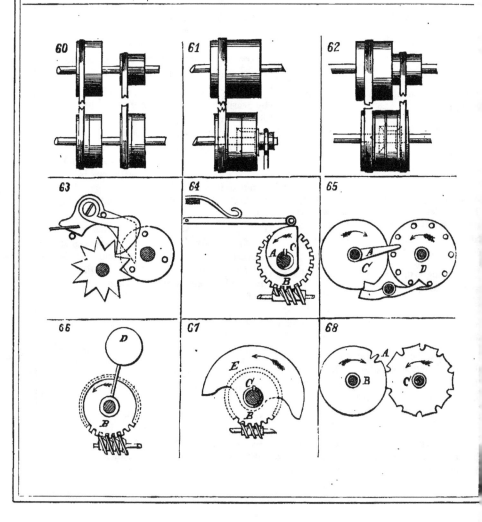

60. For transmitting two speeds by means of belts. There are four pulleys on the lower shaft, the two outer ones being loose and the two inner ones fast. The band to the left is shown on its loose pulley, the one to the right on its fast one ; a slow motion is consequently transmitted to lower shaft. When band to the right is moved on to its loose pulley, and left-hand one on to its fast pulley, a quicker motion is transmitted.

61. For transmitting two speeds, one a differential motion. The band is shown on a loose pulley on lower shaft. The middle pulley is fast on said shaft, and has a small bevel-gear secured to its hub. Pulley on the right, which, like that on the left, is loose on shaft, carries, transversely, another bevel-gear. A third bevel-gear, loose upon the shaft, is held by a friction-band which is weighted at the end. On moving band on middle pulley a simple motion is the result, but when it is moved to right-hand pulley a double speed is given to shaft. The friction-band or curb on the third bevel-gear is to allow it to slip a little on a sudden change of speed.

62. For transmitting two speeds, one of which is a different and variable motion. This is very similar to the last, except in the third bevel-gear being attached to a fourth pulley, at the right of the other three, and driven by a band from a small pulley on shaft above. When left-hand belt is on the pulley carrying the middle bevel-gear, and pulley at the right turns in the same direction, the amount of rotation of the third bevel-gear must be deducted from the double speed which the shaft would have if this gear was at rest. If, on the contrary, the right-hand belt be crossed so as to turn the pulley in an opposite direction, that amount must be added.

63. Jumping or intermittent rotary motion, used for meters and revolution-counters. The drop and attached pawl, carried by a spring at the left, are lifted by pins in the disk at the right. Pins escape first from pawl, which drops into next space of the star-wheel. When pin escapes from drop, spring throws down suddenly the drop, the pin on which strikes the pawl, which, by its action on star-wheel, rapidly gives it a portion of a revolution. This is repeated as each pin passes.

64. Another arrangement of jumping motion. Motion is communicated to worm-gear, B, by worm or endless screw at the bottom, which is fixed upon the driving-shaft. Upon the shaft carrying the worm-gear works another hollow shaft, on which is fixed cam, A. A short piece of this hollow shaft is half cut away. A pin fixed in worm-gear shaft turns hollow shaft and cam, the spring which presses on cam holding hollow shaft back against the pin until it arrives a little further than shown in the figure, when, the direction of the pressure being changed by the peculiar shape of cam, the latter falls down suddenly, independently of worm-wheel, and remains at rest till the pin overtakes it, when the same action is repeated.

65. The left-hand disk or wheel, C, is the driving-wheel, upon which is fixed the tappet, A. The other disk or wheel, D, has a series of equi-distant studs projecting from its face. Every rotation of the tappet acting upon one of the studs in the wheel, D, causes the latter wheel to move the distance of one stud. In order that this may not be exceeded, a lever-like stop is arranged on a fixed center. This stop operates in a notch cut in wheel, C, and at the instant tappet, A, strikes a stud, said notch faces the lever. As wheel, D, rotates, the end between studs is thrust out, and the other extremity enters the notch ; but immediately on the tappet leaving stud, the lever is again forced up in front of next stud, and is there held by periphery of C pressing on its other end.

66. A modification of 64 ; a weight, D, attached to an arm secured in the shaft of the worm-gear, being used instead of spring and cam.

67. Another modification of 64 ; a weight or tumbler, E, secured on the hollow shaft, being used instead of spring and cam, and operating in combination with pin, C, in the shaft of worm-gear.

68. The single tooth, A, of the driving-wheel, B, acts in the notches of the wheel, C, and turns the latter the distance of one notch in every revolution of C. No stop is necessary in this movement, as the driving-wheel, B, serves as a lock by fitting into the hollows cut in the circumference of the wheel, C, between its notches.

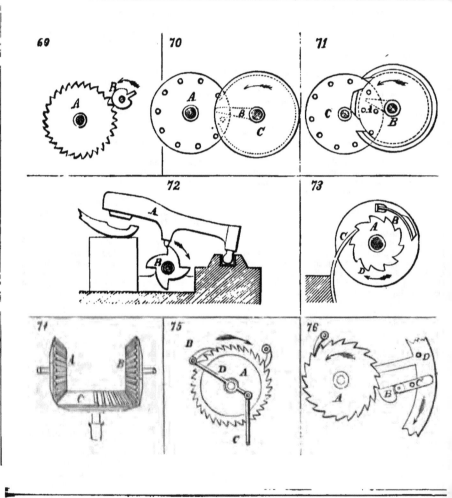

69. B, a small wheel with one tooth, is the driver, and the circumference entering between the teeth of the wheel, A, serves as a lock or stop while the tooth of the small wheel is out of operation.

70. The driving-wheel, C, has a rim, shown in dotted outline, the exterior of which serves as a bearing and stop for the studs on the other wheel, A, when the tappet, B, is out of contact with the studs. An opening in this rim serves to allow one stud to pass in and another to pass out. The tappet is opposite the middle of this opening.

71. The inner circumference (shown by dotted lines) of the rim of the driving-wheel, B, serves as a lock against which two of the studs in the wheel, C, rest until the tappet, A, striking one of the studs, the next one below passes out from the guard-rim through the lower notch, and another stud enters the rim through the upper notch.

72. Is a tilt-hammer motion, the revolution of the cam or wiper-wheel, B, lifting the hammer, A, four times in each revolution.

73. To the driving-wheel, D, is secured a bent spring, B ; another spring, C, is attached to a fixed support. As the wheel, D, revolves, the spring, B, passes under the strong spring, C, which presses it into a tooth of the ratchet-wheel, A, which is thus made to rotate. The catch-spring, B, being released on its escape from the strong spring, C, allows the wheel, A, to remain at rest till D has made another revolution. The spring, C, serves as a stop.

74. A uniform intermittent rotary motion in opposite directions is given to the bevel-gears, A and B, by means of the mutilated bevel-gear, C.

75. Reciprocating rectilinear motion of the rod, C, transmits an intermittent circular motion to the wheel, A, by means of the pawl, B, at the end of the vibrating-bar, D.

76. Is another contrivance for registering or counting revolutions. A tappet, B, supported on the fixed pivot, C, is struck at every revolution of the large wheel (partly represented) by a stud, D, attached to the said wheel. This causes the end of the tappet next the ratchet-wheel, A, to be lifted, and to turn the wheel the distance of one tooth. The tappet returns by its own weight to its original position after the stud, D, has passed, the end being jointed to permit it to pass the teeth of the ratchet-wheel.

77

78

79

80

81

82

83

84

85

77. The vibration of the lever, C, on the center or fulcrum, A, produces a rotary movement of the wheel, B, by means of the two pawls, which act alternately. This is almost a continuous movement.

78. A modification of 77.

79. Reciprocating rectilinear motion of the rod, B, produces a nearly continuous rotary movement of the ratchet-faced wheel, A, by the pawls attached to the extremities of the vibrating radial arms, C, C.

80. Rectilinear motion is imparted to the slotted bar, A, by the vibration of the lever, C, through the agency of the two hooked pawls, which drop alternately into the teeth of the slotted rack-bar, A.

81. Alternate rectilinear motion is given to the rack-rod, B, by the continuous revolution of the mutilated spur-gear, A, the spiral spring, C, forcing the rod back to its original position on the teeth of the gear, A, quitting the rack.

82. On motion being given to the two treadles, D, a nearly continuous motion is imparted, through the vibrating arms, B, and their attached pawls, to the ratchet-wheel, A. A chain or strap attached to each treadle passes over the pulley, C, and as one treadle is depressed the other is raised.

83. A nearly continuous rotary motion is given to the wheel, D, by two ratchet-toothed arcs, C, one operating on each side of the ratchet-wheel, D. These arcs (only one of which is shown) are fast on the same rock-shaft, B, and have their teeth set opposite ways. The rock-shaft is worked by giving a reciprocating rectilinear motion to the rod, A. The arcs should have springs applied to them, so that each may be capable of rising to allow its teeth to slide over those of the wheel in moving one way.

84. The double rack-frame, B, is suspended from the rod, A. Continuous rotary motion is given to the cam, D. When the shaft of the cam is midway between the two racks, the cam acts upon neither of them; but by raising or lowering the rod, A, either the lower or upper rack is brought within range of the cam, and the rack-frame moved to the left or right. This movement has been used in connection with the governor of an engine, the rod, A, being connected with the governor, and the rack-frame with the throttle or regulating valve.

85. Intermittent alternating rectilinear motion is given to the rod, A, by the continuous rotation of the shaft carrying the two cams or wipers, which act upon the projection, B, of the rod, and thereby lift it. The rod drops by its own weight. Used for ore-stampers or pulverizers, and for hammers.

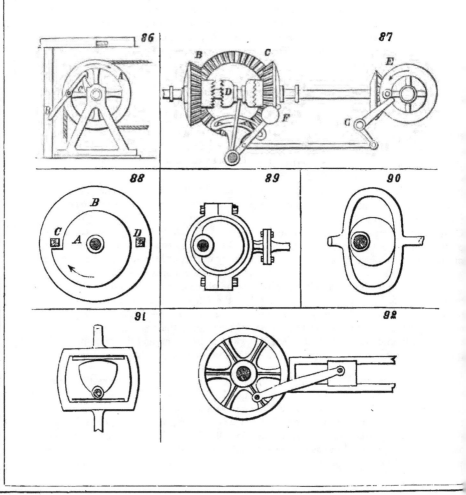

86. A method of working a reciprocating pump by rotary motion. A rope, carrying the pump-rod, is attached to the wheel, A, which runs loosely upon the shaft. The shaft carries a cam, C, and has a continuous rotary motion. At every revolution the cam seizes the hooked catch, B, attached to the wheel, and drags it round, together with the wheel, and raises the rope until, on the extremity of the catch striking the stationary stop above, the catch is released, and the wheel is returned by the weight of the pump-bucket.

87. A contrivance for a self-reversing motion. The bevel-gear between the gears, B and C, is the driver. The gears, B and C, run loose upon the shaft, consequently motion is only communicated when one or other of them is engaged with the clutch-box, D, which slides on a feather on the shaft and is shown in gear with C. The wheel, E, at the right, is driven by bevel-gearing from the shaft on which the gears, B, C, and clutch are placed, and is about to strike the bell-crank, G, and produce such a movement thereof as will cause the connecting-rod to carry the weighted lever, F, beyond a perpendicular position, when the said lever will fall over suddenly to the left, and carry the clutch into gear with B, thereby reversing the motion of the shaft, until the stud in the wheel, E, coming round in the contrary direction, brings the weighted lever back past the perpendicular position, and thereby again causes it to reverse the motion.

88. Continuous rotary converted into intermittent rotary motion. The disk-wheel, B, carrying the stops, C, D, turns on a center eccentric to the cam, A. On continuous rotary motion being given to the cam, A, intermittent rotary motion is imparted to the wheel, B. The stops free themselves from the offset of the cam at every half-revolution, the wheel, B, remaining at rest until the cam has completed its revolution, when the same motion is repeated.

89. An eccentric generally used on the crank-shaft for communicating the reciprocating rectilinear motion to the valves of steam engines, and sometimes used for pumping.

90. A modification of the above; an elongated yoke being substituted for the circular strap, to obviate the necessity for any vibrating motion of the rod which works in fixed guides.

91. Triangular eccentric, giving an intermittent reciprocating rectilinear motion, used in France for the valve motion of steam engines.

92. Ordinary crank motion.

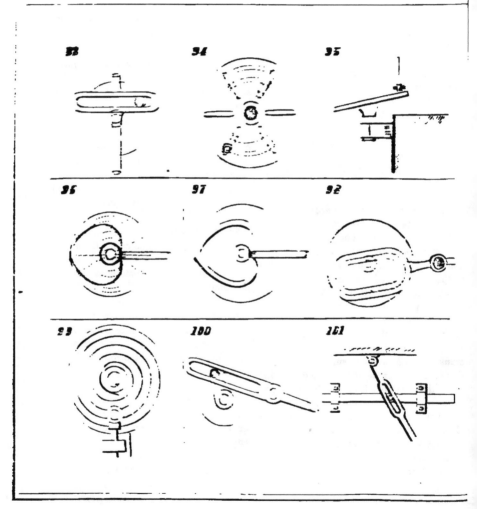

93. Crank motion, with the crank-wrist working in a slotted yoke, thereby dispensing with the oscillating connecting-rod or pitman.

94. Variable crank, two circular plates revolving on the same center. In one a spiral groove is cut; in the other a series of slots radiating from the center. On turning one of these plates around its center, the bolt shown near the bottom of the figure, and which passes through the spiral groove and radial slots, is caused to move toward or from the center of the plates.

95. On rotating the upright shaft, reciprocating rectilinear motion is imparted by the oblique disk to the upright rod resting upon its surface.

96. A heart-cam. Uniform traversing motion is imparted to the horizontal bar by the rotation of the heart-shaped cam. The dotted lines show the mode of striking out the curve of the cam. The length of traverse is divided into any number of parts; and from the center a series of concentric circles are described through these points. The outside circle is then divided into double the number of these divisions, and lines drawn to the center. The curve is then drawn through the intersections of the concentric circles and the radiating lines.

97. This is a heart-cam, similar to 96, except that it is grooved.

98. Irregular vibrating motion is produced by the rotation of the circular disk, in which is fixed a crank-pin working in an endless groove cut in the vibrating arm.

99. Spiral guide attached to the face of a disk; used for the feed-motion of a drilling machine.

100. Quick return crank motion, applicable to shaping machines.

101. Rectilinear motion of horizontal bar, by means of vibrating slotted bar hung from the top.

102

103

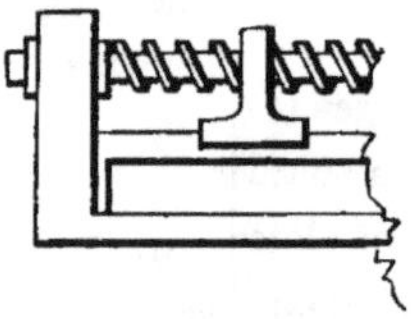

104

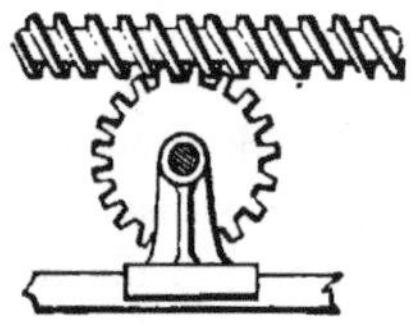

105

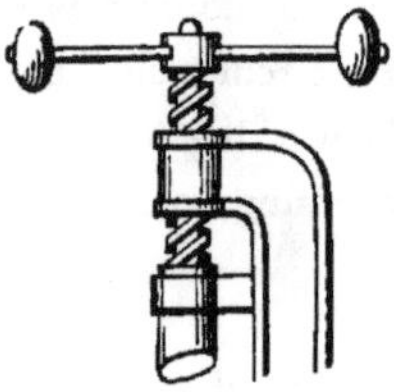

106

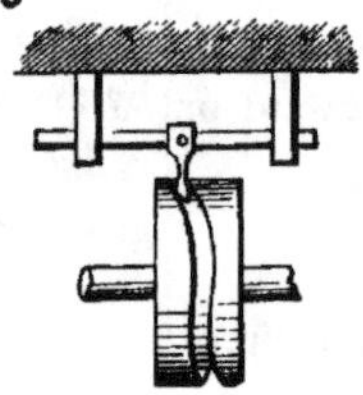

107

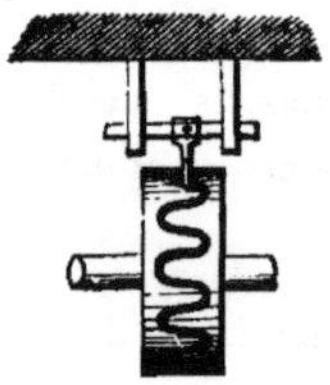

108

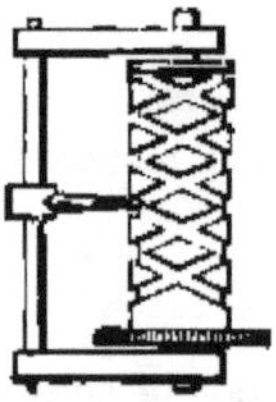

109

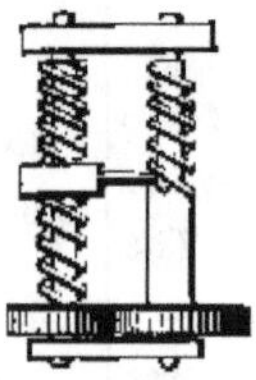

110

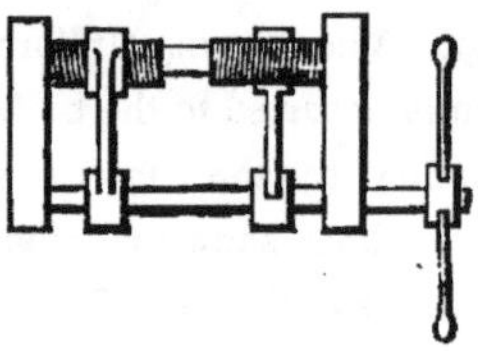

102. Common screw bolt and nut ; rec-linear motion obtained from circular mo-on.

103. Rectilinear motion of slide produced y the rotation of screw.

104. In this, rotary motion is imparted to ae wheel by the rotation of the screw, or ectilinear motion of the slide by the rota-on of the wheel. Used in screw-cutting nd slide-lathes.

105. Screw stamping-press. Rectilinear aotion from circular motion.

106 and 107. Uniform reciprocating rec-linear motion, produced by rotary motion f grooved cams.

108. Uniform reciprocating rectilinear mo-on from uniform rotary motion of a cylin-er, in which are cut reverse threads or rooves, which necessarily intersect twice

in every revolution. A point inserted in the groove will traverse the cylinder from end to end.

109. The rotation of the screw at the left-hand side produces a uniform rectilinear movement of a cutter which cuts another screw thread. The pitch of the screw to be cut may be varied by changing the sizes of the wheels at the end of the frame.

110. Uniform circular into uniform recti-linear motion ; used in spooling-frames for leading or guiding the thread on to the spools. The roller is divided into two parts, each having a fine screw thread cut upon it, one a right and the other a left hand screw. The spindle parallel with the roller has arms which carry two half-nuts, fitted to the screws, one over and the other under the roller. When one half-nut is in, the other is out of gear. By pressing the lever to the right or left, the rod is made to traverse in either direction.

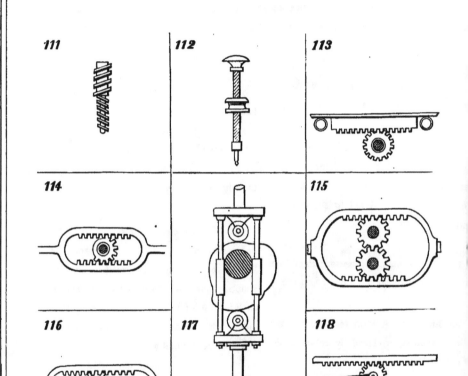

111. Micrometer screw. Great power can be obtained by this device. The threads are made of different pitch and run in different directions, consequently a die or nut fitted to the inner and smaller screw would traverse only the length of the difference between the pitches for every revolution of the outside hollow screw in a nut.

112. Persian drill. The stock of the drill has a very quick thread cut upon it and revolves freely, supported by the head at the top, which rests against the body. The button or nut shown on the middle of the screw is held firm in the hand, and pulled quickly up and down the stock, thus causing it to revolve to the right and left alternately.

113. Circular into rectilinear motion, or *vice versa*, by means of rack and pinion.

114. Uniform circular motion into reciprocating rectilinear motion, by means of mutilated pinion, which drives alternately the top and bottom rack.

115. Rotary motion of the toothed wheels produces rectilinear motion of the double rack and gives equal force and velocity to each side, both wheels being of equal size.

116. A substitute for the crank. Reciprocating rectilinear motion of the frame carrying the double rack produces a uniform rotary motion of the pinion-shaft. A separate pinion is used for each rack, the two racks being in different planes. Both pinions are loose on the shaft. A ratchet-wheel is fast on the shaft outside of each pinion, and a pawl attached to the pinion to engage in it, one ratchet-wheel having its teeth set in one direction and the other having its teeth set in the opposite direction. When the racks move one way, one pinion turns the shaft by means of its pawl and ratchet; and when the racks move the opposite way, the other pinion acts in the same way, one pinion always turning loosely on the shaft.

117. A cam acting between two friction-rollers in a yoke. Has been used to give the movement to the valve of a steam engine.

118. A mode of doubling the length of stroke of a piston-rod, or the throw of a crank. A pinion revolving on a spindle attached to the connecting-rod or pitman is in gear with a fixed rack. Another rack carried by a guide-rod above, and in gear with the opposite side of the pinion, is free to traverse backward and forward. Now, as the connecting-rod communicates to the pinion the full length of stroke, it would cause the top rack to traverse the same distance, if the bottom rack was alike movable; but as the latter is fixed, the pinion is made to rotate, and consequently the top rack travels double the distance.

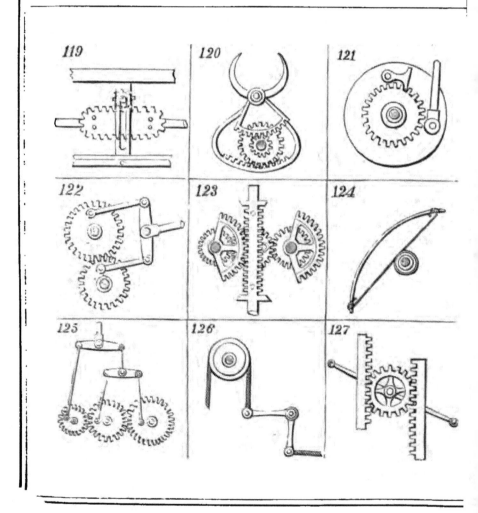

119. Reciprocating rectilinear motion of the bar carrying the oblong endless rack, produced by the uniform rotary motion of the pinion working alternately above and below the rack. The shaft of the pinion moves up and down in, and is guided by, the slotted bar.

120. Each jaw is attached to one of the two segments, one of which has teeth outside and the other teeth inside. On turning the shaft carrying the two pinions, one of which gears with one and the other with the other segment, the jaws are brought together with great force.

121. Alternating rectilinear motion of the rod attached to the disk-wheel produces an intermittent rotary motion of the cog-wheel by means of the click attached to the disk-wheel. This motion, which is reversible by throwing over the click, is used for the feed of planing machines and other tools.

122. The rotation of the two spur-gears, with crank-wrists attached, produces a variable alternating traverse of the horizontal bar.

123. Intended as a substitute for the crank. Reciprocating rectilinear motion of the double rack gives a continuous rotary motion to the center gear. The teeth on the rack act upon those of the two semi-circular toothed sectors, and the spur-gears attached to the sectors operate upon the center gear. The two stops on the rack shown by dotted lines are caught by the curved piece on the center gear, and lead the toothed sectors alternately into gear with the double rack.

124. Fiddle drill. Reciprocating rectilinear motion of the bow, the string of which passes around the pulley on the spindle carrying the drill, producing alternating rotary motion of the drill.

125. A modification of the motion shown in 122, but of a more complex character.

126. A bell-crank lever, used for changing the direction of any force.

127. Motion used in air-pumps. On vibrating the lever fixed on the same shaft with the spur-gear, reciprocating rectilinear motion is imparted to the racks on each side, which are attached to the pistons of two pumps, one rack always ascending while the other is descending.

128. A continuous rotary motion of the shaft carrying the three wipers produces a reciprocating rectilinear motion of the rectangular frame. The shaft must revolve in the direction of the arrow for the parts to be in the position represented.

129. Chinese windlass. This embraces the same principles as the micrometer screw 111. The movement of the pulley in every revolution of the windlass is equal to half the difference between the larger and smaller circumferences of the windlass barrel.

130. Shears for cutting iron plates, etc. The jaws are opened by the weight of the long arm of the upper one, and closed by the rotation of the cam.

131. On rotating the disk carrying the crank-pin working in the slotted arm, reciprocating rectilinear motion is imparted to the rack at the bottom by the vibration of the toothed sector.

132. This is a motion which has been used in presses to produce the necessary pressure upon the platen. Horizontal motion is given to the arm of the lever which turns the upper disk. Between the top and bottom disks are two bars which enter holes in the disks. These bars are in oblique positions, as shown in the drawing, when the press is not in operation; but when the top disk is made to rotate, the bars move toward perpendicular positions and force the lower disk down. The top disk must be firmly secured in a stationary position, except as to its revolution.

133. A simple press motion is given through the hand-crank on the pinion-shaft; the pinion communicating motion to the toothed sector, which acts upon the platen, by means of the rod which connects it therewith.

134. Uniform circular motion into rectilinear by means of a rope or band, which is wound once or more times around the drum.

135. Modification of the triangular eccentric 91, used on the steam engine in the Paris Mint. The circular disk behind carries the triangular tappet, which communicates an alternate rectilinear motion to the valve-rod. The valve is at rest at the completion of each stroke for an instant, and is pushed quickly across the steam-ports to the end of the next.

136. A cam-wheel — of which a side view is shown — has its rim formed into teeth, or made of any profile form desired. The rod to the right is made to press constantly against the teeth or edge of the rim. On turning the wheel, alternate-rectilinear motion is communicated to the rod. The character of this motion may be varied by altering the shape of the teeth or profile of the edge of the rim of the wheel.

137

138

139

140

141

142

143

144

145

137. Expansion eccentric used in France to work the slide-valve of a steam engine. The eccentric is fixed on the crank-shaft, and communicates motion to the forked vibrating arm to the bottom of which the valve-rod is attached.

138. On turning the cam at the bottom a variable alternating rectilinear motion is imparted to the rod resting on it.

139. The internal rack, carried by the rectangular frame, is free to slide up and down within it for a certain distance, so that the pinion can gear with either side of the rack. Continuous circular motion of the pinion is made to produce reciprocating rectilinear motion of rectangular frame.

140. The toggle-joint arranged for a punching machine. Lever at the right is made to operate upon the joint of the toggle by means of the horizontal connecting-link.

141. Endless-band saw. Continuous rotary motion of the pulleys is made to produce continuous rectilinear motion of the straight parts of the saw.

142. Movement used for varying the length of the traversing guide-bar which, in silk machinery, guides the silk on to spools or bobbins. The spur-gear, turning freely on its center, is carried round by the larger circular disk, which turns on a fixed central stud, which has a pinion fast on its end. Upon the spur-gear is bolted a small crank, to which is jointed a connecting-rod attached to traversing guide-bar. On turning the disk, the spur-gear is made to rotate partly upon its center by means of the fixed pinion, and consequently brings crank nearer to center of disk. If the rotation of disk was continued, the spur-gear would make an entire revolution. During half a revolution the traverse would have been shortened a certain amount at every revolution of disk, according to the size of spur-gear; and during the other half it would have gradually lengthened in the same ratio.

143. Circular motion into alternate rectilinear motion. Motion is transmitted through pulley at the left upon the worm-shaft. Worm slides upon shaft, but is made to turn with it by means of a groove cut in shaft, and a key in hub of worm. Worm is carried by a small traversing-frame, which slides upon a horizontal bar of the fixed frame, and the traversing-frame also carries the toothed wheel into which the worm gears. One end of a connecting-rod is attached to fixed frame at the right and the other end to a wrist secured in toothed wheel. On turning worm-shaft, rotary motion is transmitted by worm to wheel, which, as it revolves, is forced by connecting-rod to make an alternating traverse motion.

144. A system of crossed levers, termed "Lazy Tongs." A short alternating rectilinear motion of rod at the right will give a similar but much greater motion to rod at the left. It is frequently used in children's toys. It has been applied in France to a machine for raising sunken vessels; also applied to ships' pumps, three-quarters of a century ago.

145. Reciprocating curvilinear motion of the beam gives a continuous rotary motion to the crank and fly-wheel. The small standard at the left, to which is attached one end of the lever with which the beam is connected by the connecting-rod, has a horizontal reciprocating rectilinear movement.

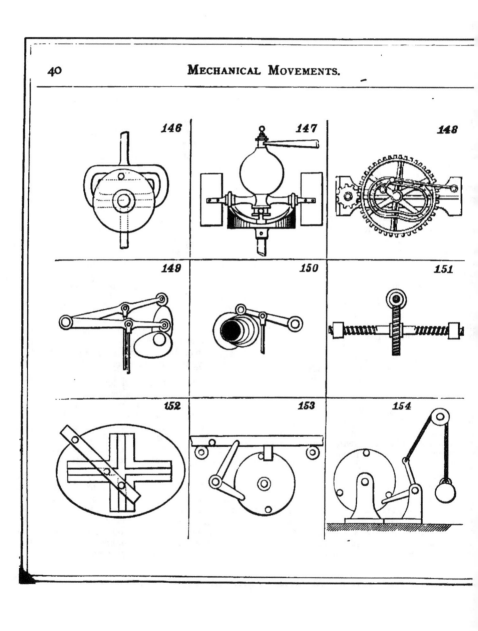

146. Continuous rotary motion of the disk produces reciprocating rectilinear motion of the yoke-bar, by means of the wrist or crank-pin on the disk working in the groove of the yoke. The groove may be so shaped as to obtain a uniform reciprocating rectilinear motion.

147. Steam engine governor. The operation is as follows:— On engine starting the spindle revolves and carries round the cross-head to which fans are attached, and on which are also fitted two friction-rollers which bear on two circular inclined planes attached securely to the center shaft, the cross-head being loose on the shaft. The cross-head is made heavy, or has a ball or other weight attached, and is driven by the circular inclined planes. As the speed of the center shaft increases, the resistance of the air to the wings tends to retard the rotation of the cross-head; the friction-rollers therefore run up the inclined planes and raise the cross-head, to the upper part of which is connected a lever operating upon the regulating-valve of the engine.

148. Continuous circular motion of the spur-gears produces alternate circular motion of the crank attached to the larger gear.

149. Uniform circular converted, by the cams acting upon the levers, into alternating rectilinear motions of the attached rods.

150. A valve motion for working steam expansively. The series of cams of varying throw are movable lengthwise of the shaft so that either may be made to act upon the lever to which the valve-rod is connected. A greater or less movement of the valve is produced, according as a cam of greater or less throw is opposite the lever.

151. Continuous circular into continuous but much slower rectilinear motion. The worm on the upper shaft, acting on the toothed wheel on the screw-shaft, causes the right and left hand screw-threads to move the nuts upon them toward or from each other according to the direction of rotation.

152. An ellipsograph. The traverse bar (shown in an oblique position) carries two studs which slide in the grooves of the cross-piece. By turning the traverse bar an attached pencil is made to describe an ellipse by the rectilinear movement of the studs in the grooves.

153. Circular motion into alternating rectilinear motion. The studs on the rotating disk strike the projection on the under side of the horizontal bar, moving it one direction. The return motion is given by means of the bell-crank or elbow-lever, one arm of which is operated upon by the next stud, and the other strikes the stud on the front of the horizontal bar.

154. Circular motion into alternating rectilinear motion, by the action of the studs on the rotary disk upon one end of the bell-crank, the other end of which has attached to it a weighted cord passing over a pulley.

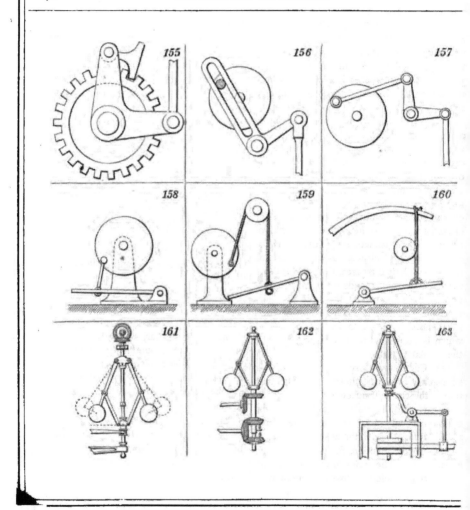

155. Reciprocating rectilinear motion to intermittent circular motion by means of the pawl attached to the elbow-lever, and operating in the toothed wheel. Motion is given to the wheel in either direction according to the side on which the pawl works. This is used in giving the feed-motion to planing machines and other tools.

156. Circular motion into variable alternating rectilinear motion, by the wrist or crank-pin on the rotating disk working in the slot of the bell-crank or elbow-lever.

157. A modification of the movement last described ; a connecting-rod being substituted for the slot in the bell-crank.

158. Reciprocating curvilinear motion of the treadle gives a circular motion to the disk. A crank may be substituted for the disk.

159. A modification of 158, a cord and pulley being substituted for the connecting rod.

160. Alternating curvilinear motion into alternating circular. When the treadle has been depressed, the spring at the top elevates it for the next stroke ; the connecting band passes once round the pulley, to which gives motion.

161. Centrifugal governor for steam engines. The central spindle and attached arms and balls are driven from the engine by the bevel-gears at the top, and the balls fly out from the center by centrifugal force. If the speed of the engine increases, the balls fly out further from the center, and so raise the slide at the bottom and thereby reduce the opening of the regulating-valve which is connected with said slide. A diminution of speed produces an opposite effect.

162. Water-wheel governor acting on the same principle as 161, but by different means. The governor is driven by the top horizontal shaft and bevel-gears, and the lower gears control the rise and fall of the shuttle or gate over or through which the water flows to the wheel. The action is as follows :—The two bevel-gears on the lower part of the center spindle, which are furnished with studs, are fitted loosely to the said spindle and remain at rest so long as the governor has a proper velocity ; but immediately that the velocity increases, the balls, flying further out, draw up the pin which is attached to a loose sleeve which slides up and down the spindle, and this pin, coming in contact with the stud on the upper bevel gear, causes that gear to rotate with the spindle and to give motion to the lower horizontal shaft in such a direction as to make it raise the shuttle or gate, and so reduce the quantity of water passing to the wheel. On the contrary, if the speed of the governor decreases below that required, the pin falls and gives motion to the lower bevel-gear, which drives the horizontal shaft in the opposite direction and produces a contrary effect.

163. Another arrangement for a water-wheel governor. In this the governor controls the shuttle or gate by means of the cranked lever, which acts on the strap or belt in the following manner :—The belt runs on one of three pulleys, the middle one of which is loose on the governor spindle and the upper and lower ones fast. When the governor is running at the proper speed the belt is on the loose pulley, as shown ; but when the speed increases the belt is thrown on the lower pulley, and thereby caused to act upon suitable gearing for raising the gate or shuttle and decreasing the supply of water. A reduction of the speed of the governor brings the belt on the upper pulley, which acts upon gearing for producing an opposite effect on the shuttle or gate.

164

165

166

167

168

171

169

170

TRUNNION

164. A knee-lever, differing slightly from the toggle-joint shown in 40. It is often used for presses and stamps, as a great force can be obtained by it. The action is by raising or lowering the horizontal lever.

165. Circular into rectilinear motion. The waved-wheel or cam on the upright shaft communicates a rectilinear motion to the upright bar through the oscillating rod.

166. The rotation of the disk carrying the crank pin gives a to-and-fro motion to the connecting-rod, and the slot allows the rod to remain at rest at the termination of each stroke ; it has been used in a brick-press, in which the connecting-rod draws a mold backward and forward, and permits it to rest at the termination of each stroke, that the clay may be deposited in it and the brick extracted.

167. A drum or cylinder having an endless spiral groove extending all around it ; one half of the groove having its pitch in one, and the other half its pitch in the opposite direction. A stud on a reciprocating rectilinearly moving rod works in the groove, and so converts reciprocating into rotary motion. This has been used as a substitute for the crank in a steam engine.

168. The slotted crank at the left hand of the figure is on the main shaft of an engine, and the pitman which connects it with the reciprocating moving power is furnished with a pin which works in the slot of the crank. Intermediate between the first crank and the moving power is a shaft carrying a second crank, of an invariable radius, connected with the same pitman. While the first crank moves in a circular orbit, the pin at the end of the pitman is compelled to move in an elliptical orbit, thereby increasing the leverage of the main crank at those points which are most favorable for the transmission of power.

169. A modification of 168, in which a link is used to connect the pitman with the main crank, thereby dispensing with the slot in the said crank.

170. Another form of steam engine governor. Instead of the arms being connected with a slide working on a spindle, they cross each other and are elongated upward beyond the top thereof and connected with the valve-rod by two short links.

171. Valve motion and reversing gear used in oscillating marine engines. The two eccentric rods give an oscillating motion to the slotted link which works the curved slide over the trunnion. Within the slot in the curved slide is a pin attached to the arm of a rock-shaft which gives motion to the valve. The curve of the slot in the slide is an arc of a circle described from the center of the trunnion, and as it moves with the cylinder it does not interfere with the stroke of the valve. The two eccentrics and link are like those of the link motion used in locomotives.

172

173

174

175

176

177

178

179

180

172. A mode of obtaining an egg-shaped elliptical movement.

173. A movement used in silk machinery for the same purpose as that described in 142. On the back of a disk or bevel-gear is secured a screw with a tappet-wheel at one extremity. On each revolution of the disk the tappet-wheel comes in contact with a pin or tappet, and thus receives an intermittent rotary movement. A wrist secured to a nut on the screw enters and works in a slotted bar at the end of the rod which guides the silk on the bobbins. Each revolution of the disk varies the length of stroke of the guide-rod, as the tappet-wheel on the end of the screw turns the screw with it, and the position of the nut on the screw is therefore changed.

174. Carpenters' bench-clamp. By pushing the clamp between the jaws they are made to turn on the screws and clamp the sides.

175. A means of giving one complete revolution to the crank of an engine to each stroke of the piston.

176 and 177. Contrivance for uncoupling engines. The wrist which is fixed on one arm of the crank (not shown) will communicate motion to the arm of the crank which is represented, when the ring on the latter has its slot in the position shown in 176. But when the ring is turned to bring the slot in the position shown in 177, the wrist passes through the slot without turning the crank to which said ring is attached.

178. Contrivance for varying the speed of the slide carrying the cutting tool in slotting and shaping machines, etc. The driving-shaft works through an opening in a fixed disk, in which is a circular slot. At the end of the said shaft is a slotted crank. A slide fits in the slot of the crank and in the circular slot; and to the outward extremity of this slide is attached the connecting-rod which works the slide carrying the cutting tool. When the driving-shaft rotates the crank is carried round, and the slide carrying the end of the connecting-rod is guided by the circular slot, which is placed eccentrically to the shaft; therefore, as the slide approaches the bottom, the length of the crank is shortened and the speed of the connecting-rod is diminished.

179. Reversing-gear for a single engine. On raising the eccentric-rod the valve-spindle is released. The engine can then be reversed by working the upright lever, after which the eccentric-rod is let down again. The eccentric in this case is loose upon the shaft and driven by a projection on the shaft acting upon a nearly semi-circular projection on the side of the eccentric, which permits the eccentric to turn half-way round on the shaft on reversing the valves.

180. This only differs from 174 in being composed of a single pivoted clamp operating in connection with a fixed side-piece.

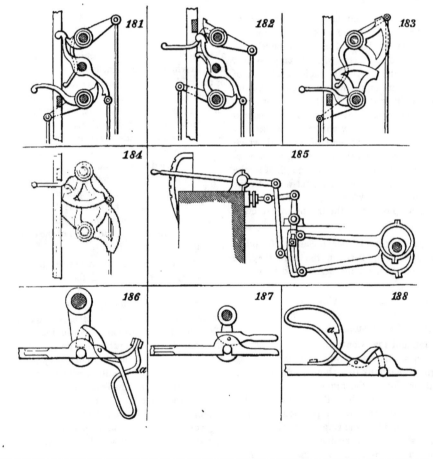

181 and 182. Diagonal catch or hand-gear used in large blowing and pumping engines. In 181 the lower steam-valve and upper eduction-valve are open, while the upper steam-valve and lower eduction-valve are shut; consequently the piston will be ascending. In the ascent of the piston-rod the lower handle will be struck by the projecting tappet, and, being raised, will become engaged by the catch and shut the upper eduction and lower steam valves; at the same time, the upper handle being disengaged from the catch, the back weight will pull the handle up and open the upper steam and lower eduction valves, when the piston will consequently descend. 182 represents the position of the catchers and handles when the piston is at the top of the cylinder. In going down, the tappet of the piston-rod strikes the upper handle and throws the catches and handles to the position shown in 181.

183 and 184 represent a modification of 181 and 182, the diagonal catches being superseded by two quadrants.

185. Link-motion valve-gear of a locomotive. Two eccentrics are used for one valve, one for the forward and the other for the backward movement of the engine. The extremities of the eccentric-rods are jointed to a curved slotted bar, or, as it is termed, a *link*, which can be raised or lowered by an arrangement of levers terminating in a handle as shown. In the slot of the link is a slide and pin connected with an arrangement of levers terminating at the valve-stem. The link, in moving with the action of the eccentrics, carries with it the slide, and thence motion is communicated to the valve. Suppose the link raised so that the slide is in the middle, then the link will oscillate on the pin of the slide, and consequently the valve will be at rest. If the link is moved so that the slide is at one of its extremities, the whole throw of the eccentric connected with that extremity will be given to it, and the valve and steam-ports will be opened to the full, and it will only be toward the end of the stroke that they will be totally shut, consequently the steam will have been admitted to the cylinder during almost the entire length of each stroke. But if the slide is between the middle and the extremity of the slot, as shown in the figure, it receives only a part of the throw of the eccentric, and the steam-ports will only be partially opened, and are quickly closed again, so that the admission of steam ceases some time before the termination of the stroke, and the steam is worked expansively. The nearer the slide is to the middle of the slot the greater will be the expansion, and *vice versa*.

186. Apparatus for disengaging the eccentric-rod from the valve-gear. By pulling up the spring handle below until it catches in the notch, *a*, the pin is disengaged from the gab in the eccentric-rod.

187 and 188. Modifications of 186.

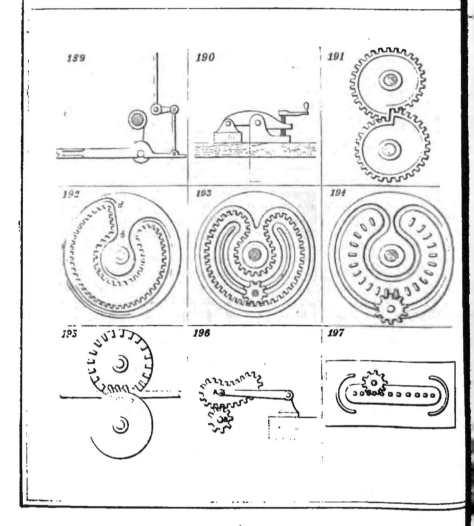

189. Another modification of 186.

190. A screw-clamp. On turning the handle the screw thrusts upward against the holder, which, operating as a lever, holds down the piece of wood or other material placed under it on the other side of its fulcrum.

191. Scroll-gears for obtaining a gradually increasing speed.

192. A variety of what is known as the "mangle-wheel." One variety of this was illustrated by 36. In this one the speed varies in every part of a revolution, the groove, *b*, *d*, in which the pinion-shaft is guided, as well as the series of teeth, being eccentric to the axis of the wheel.

193. Another kind of mangle-wheel with its pinion. With this as well as with that in the preceding figure, although the pinion continues to revolve in one direction, the mangle-wheel will make almost an entire revolution in one direction and the same in an opposite direction; but the revolution of the wheel in one direction will be slower than that in the other, owing to the greater radius of the outer circle of teeth.

194. Another mangle-wheel. In this the speed is equal in both directions of motion,

only one circle of teeth being provided on the wheel. With all of these mangle-wheels the pinion-shaft is guided and the pinion kept in gear by a groove in the wheel. The said shaft is made with a universal joint, which allows a portion of it to have the vibratory motion necessary to keep the pinion in gear.

195. A mode of driving a pair of feed-rolls, the opposite surfaces of which require to move in the same direction. The two wheels are precisely similar, and both gear into the endless screw which is arranged between them. The teeth of one wheel only are visible, those of the other being on the back or side which is concealed from view.

196. The pinion, B, rotates about a fixed axis and gives an irregular vibratory motion to the arm carrying the wheel, A.

197. What is called a "mangle-rack." A continuous rotation of the pinion will give a reciprocating motion to the square frame. The pinion-shaft must be free to rise and fall, to pass round the guides at the ends of the rack. This motion may be modified as follows :—If the square frame be fixed, and the pinion be fixed upon a shaft made with a universal joint, the end of the shaft will describe a line, similar to that shown in the drawing, around the rack.

198

199

200

201

202

203

204

205

206

198. A modification of 197. In this the pinion revolves, but does not rise and fall as in the former figure. The portion of the frame carrying the rack is jointed to the main portion of the frame by rods, so that when the pinion arrives at the end it lifts the rack by its own movement, and follows on the other side.

199. Another form of mangle-rack. The lantern-pinion revolves continuously in one direction, and gives reciprocating motion to the square frame, which is guided by rollers or grooves. The pinion has only teeth in less than half of its circumference, so that while it engages one side of the rack, the toothless half is directed against the other. The large tooth at the commencement of each rack is made to insure the teeth of the pinion being properly in gear.

200. A mode of obtaining two different speeds on the same shaft from one driving-wheel.

201. A continual rotation of the pinion (obtained through the irregular shaped gear at the left) gives a variable vibrating move-ment to the horizontal arm, and a variable reciprocating movement to the rod, A.

202. Worm or endless screw and worm-wheel. Modification of 30, used when steadiness or great power is required.

203. A regular vibrating movement of the curved slotted arm gives a variable vibration to the straight arm.

204. An illustration of the transmission of rotary motion from one shaft to another, arranged obliquely to it, by means of rolling contact.

205. Represents a wheel driven by a pinion of two teeth. The pinion consists in reality of two cams, which gear with two distinct series of teeth on opposite sides of the wheel, the teeth of one series alternating in position with those of the other.

206. A continuous circular movement of the ratchet-wheel, produced by the vibration of the lever carrying two pawls, one of which engages the ratchet-teeth in rising and the other in falling.

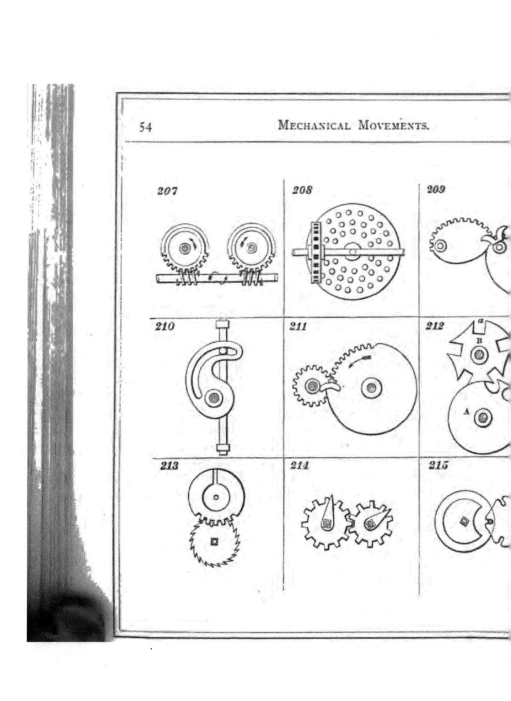

207. A modification of 195 by means of two worms and worm-wheels.

208. A pin-wheel and slotted pinion, by which three changes of speed can be obtained. There are three circles of pins of equal distance on the face of the pin-wheel, and by shifting the slotted pinion along its shaft, to bring it in contact with one or the other of the circles of pins, a continuous rotary motion of the wheel is made to produce three changes of speed of the pinion, or *vice versa*.

209. Represents a mode of obtaining motion from rolling contact. The teeth are for making the motion continuous, or it would cease at the point of contact shown in the figure. The forked catch is to guide the teeth into proper contact.

210. By turning the shaft carrying the curved slotted arm, a rectilinear motion of variable velocity is given to the vertical bar.

211. A continuous rotary motion of the large wheel gives an intermittent rotary motion to the pinion-shaft. The part of the pinion shown next the wheel is cut of the same curve as the plain portion of the circumference of the wheel, and therefore serves as a lock while the wheel makes a part of a revolution, and until the pin upon the wheel strikes the guide-piece upon the pinion, when the pinion-shaft commences another revolution.

212. What is called the "Geneva-stop, used in Swiss watches to limit the number of revolutions in winding-up; the convex curved part, *a, b,* of the wheel, B, serving as the stop.

213. Another kind of stop for the same purpose.

214 and 215. Other modifications of the stop, the operations of which will be easily understood by a comparison with 212.

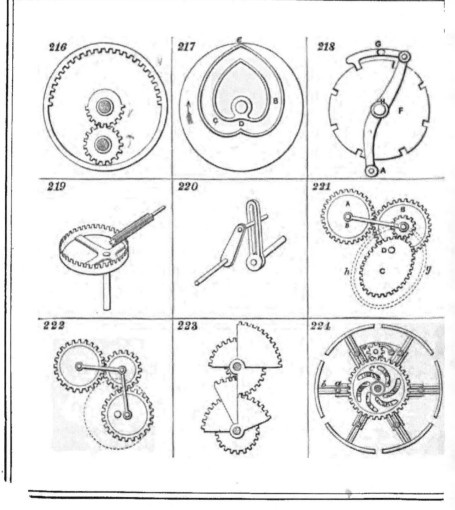

216. The external and internal mutilated :og-wheels work alternately into the pinion, und give slow forward and quick reverse motion.

217 and 218. These are parts of the same movement, which has been used for giving the roller motion in wool-combing machines. The roller to which wheel, F (218), is secured is required to make one third a revolution backward, then two thirds of a revolution forward, when it must stop until another length of combed fiber is ready for delivery. This is accomplished by the grooved heart-cam, C, D, B, e (217), the stud, A, working in the said groove; from C to D it moves the roller backward, and from D to e it moves it forward, the motion being transmitted through the catch, G, to the notch-wheel, F, on the roller-shaft, H. When the stud, A, arrives at the point, e, in the cam, a projection at the back of the wheel which carries the cam strikes the projecting piece on the catch, G, and raises it out of the notch in the wheel, F, so that, while the stud is traveling in the cam from e to C, the catch is passing over the plain surface between the two notches in the wheel, F, without imparting any motion; but when stud, A, arrives at the part, C, the catch has dropped in another notch, and is again ready to move wheel, F, and roller as required.

219. Variable circular motion by crown-wheel and pinion. The crown-wheel is placed eccentrically to the shaft, therefore the relative radius changes.

220. The two crank-shafts are parallel in direction, but not in line with each other. The revolution of either will communicate motion to the other with a varying velocity, for the wrist of one crank working in the slot of the other is continually changing its distance from the shaft of the latter.

221. Irregular circular motion imparted to wheel, A. C is an elliptical spur-gear rotating round center, D, and is the driver. B is a small pinion with teeth of the same pitch, gearing with C. The center of this pinion is not fixed, but is carried by an arm or frame which vibrates on a center, A, so that as C revolves the frame rises and falls to enable pinion to remain in gear with it, notwithstanding the variation in its radius of contact. To keep the teeth of C and B in gear to a proper depth, and prevent them from riding over each other, wheel, C, has attached to it a plate which extends beyond it and is furnished with a groove, g, h, of similar elliptical form, for the reception of a pin or small roller attached to the vibrating arm concentric with pinion, B.

222. If for the eccentric wheel described in the last figure an ordinary spur-gear moving on an eccentric center of motion be substituted, a simple link connecting the center of the wheel with that of the pinion with which it gears will maintain proper pitching of teeth in a more simple manner than the groove.

223. An arrangement for obtaining variable circular motion. The sectors are arranged on different planes, and the relative velocity changes according to the respective diameters of the sectors.

224. This represents an expanding pulley. On turning pinion, d, to the right or left, a similar motion is imparted to wheel, c, which, by means of curved slots cut therein, thrusts the studs fastened to arms of pulley outward or inward, thus augmenting or diminishing the size of the pulley.

225

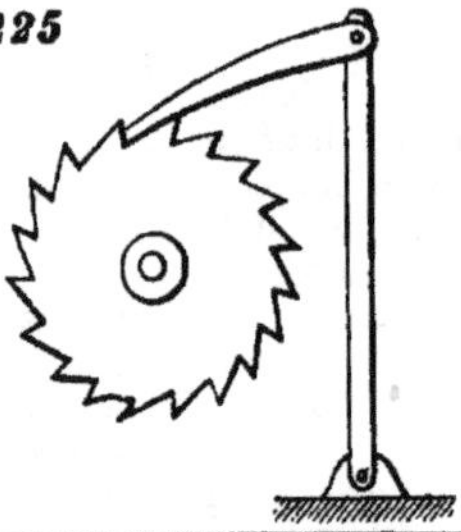

226

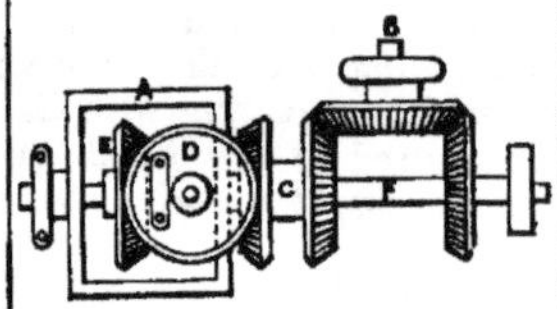

227

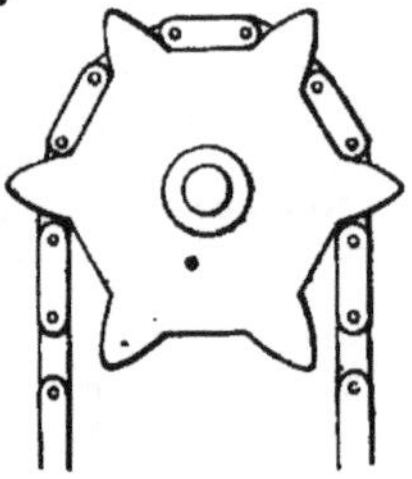

228

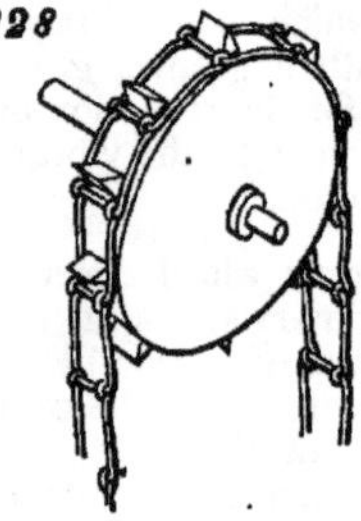

229

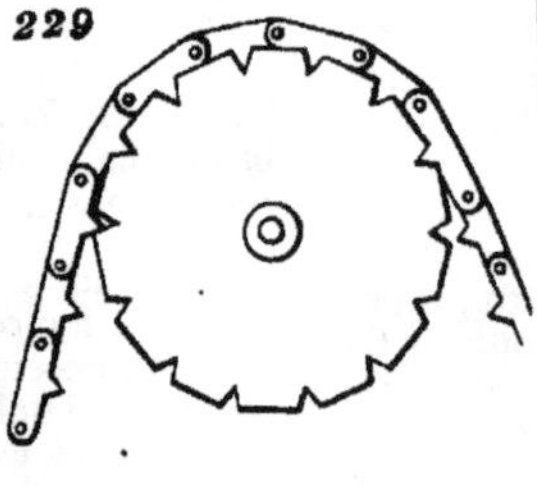

230

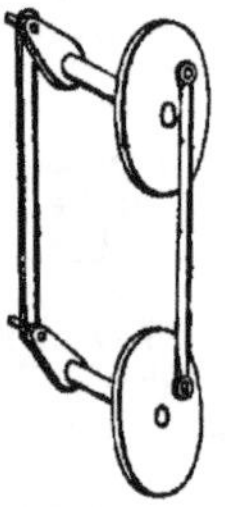

231

232

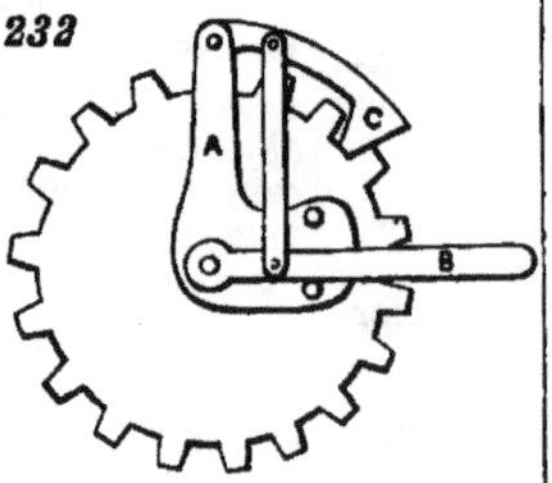

233

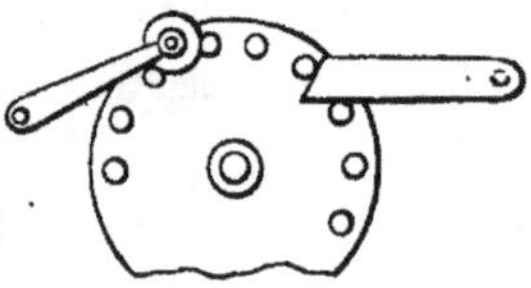

225. Intermittent circular motion of the ratchet-wheel from vibratory motion of the arm carrying a pawl.

226. This movement is designed to double the speed by gears of equal diameters and numbers of teeth—a result once generally supposed to be impossible. Six bevel-gears are employed. The gear on the shaft, B, is in gear with two others—one on the shaft, F, and the other on the same hollow shaft with C, which turns loosely on F. The gear, D, is carried by the frame, A, which, being fast on the shaft, F, is made to rotate, and therefore takes round D with it. E is loose on the shaft, F, and gears with D. Now, suppose the two gears on the hollow shaft, C, were removed and D prevented from turning on its axis; one revolution given to the gear on B would cause the frame, A, also to receive one revolution, and as this frame carries with it the gear, D, gearing with E, one revolution would be imparted to E; but if the gears on the hollow shaft, C, were replaced, D would receive also a revolution on its axis during the one revolution of B, and thus would produce two revolutions of E.

227. Represents a chain and chain pulley.

The links being in different planes, spaces are left between them for the teeth of the pulley to enter.

228. Another kind of chain and pulley.

229. Another variety.

230. Circular motion into ditto. The connecting-rods are so arranged that when one pair of connected links is over the dead point, or at the extremity of its stroke, the other is at right angles ; continuous motion is thus insured without a fly-wheel.

231. Drag-link motion. Circular motion is transmitted from one crank to the other.

232. Intermittent circular motion is imparted to the toothed wheel by vibrating the arm, B. When the arm, B, is lifted, the pawl, C, is raised from between the teeth of the wheel, and, traveling backward over the circumference, again drops between two teeth on lowering the arm, and draws with it the wheel.

233. Shows two different kinds of stops for a lantern-wheel.

234

235

236

237

238

239

240

241

242

234. Represents a verge escapement. On oscillating the spindle, S, the crown-wheel has an intermittent rotary motion.

235. The oscillation of the tappet-arm produces an intermittent rotary motion of the ratchet-wheel. The small spring at the bottom of the tappet-arm keeps the tappet in the position shown in the drawing as the arm rises, yet allows it to pass the teeth on the return motion.

236. A nearly continuous circular motion is imparted to the ratchet-wheel on vibrating the lever, *a*, to which are attached the two pawls, *b* and *c*.

237. A reciprocating circular motion of the top arm makes its attached pawl produce an intermittent circular motion of the crown-ratchet or rag-wheel.

238. An escapement. D is the escape-wheel, and C and B the pallets. A is the axis of the pallets.

239. An arrangement of stops for a spur-gear.

240. Represents varieties of stops for a ratchet-wheel.

241. Intermittent circular motion is imparted to the wheel, A, by the continuous circular motion of the smaller wheel with one tooth.

242. A brake used in cranes and hoisting machines. By pulling down the end of the lever, the ends of the brake-strap are drawn toward each other, and the strap tightened on the brake-wheel.

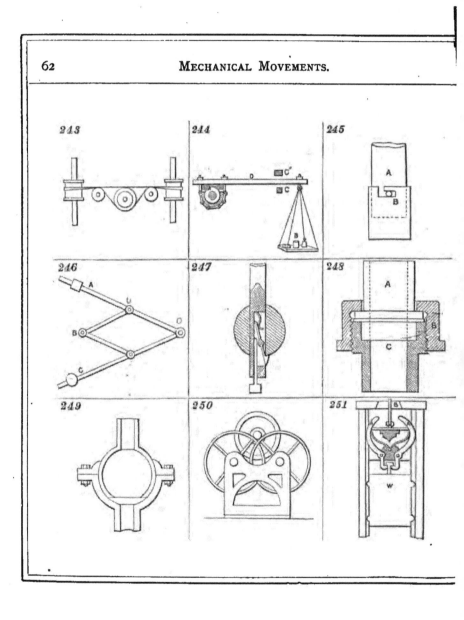

243. Represents a mode of transmitting power from a horizontal shaft to two vertical ones by means of pulleys and a band.

244. A dynamometer, or instrument used for ascertaining the amount of useful effect given out by any motive-power. It is used as follows :—A is a smoothly-turned pulley, secured on a shaft as near as possible to the motive-power. Two blocks of wood are fitted to this pulley, or one block of wood and a series of straps fastened to a band or chain, as in the drawing, instead of a common block. The blocks or block and straps are so arranged that they may be made to bite or press upon the pulley by means of the screws and nuts on the top of the lever, D. To estimate the amount of power transmitted through the shaft, it is only necessary to ascertain the amount of friction of the drum, A, when it is in motion, and the number of revolutions made. At the end of the lever, D, is hung a scale, B, in which weights are placed. The two stops, C, C', are to maintain the lever as nearly as possible in a horizontal position. Now, suppose the shaft to be in motion, the screws are to be tightened and weights added in B, until the lever takes the position shown in the drawing at the required number of revolutions. Therefore the useful effect would be equal to the product of the weights multiplied by the velocity at which the point of suspension of the weights would revolve if the lever were attached to the shaft.

245. Bayonet joint. On turning the part, A, it is released from the L-shaped slot in the socket, B, when it can be withdrawn.

246. Represents a pantograph for copying, enlarging, and reducing plans, etc. One arm is attached to and turns on the fixed point, C. B is an ivory tracing-point, and A the pencil. Arranged as shown, if we trace the lines of a plan with the point, B, the pencil will reproduce it double the size. By shifting the slide attached to the fixed point, C, and the slide carrying the pencil along their respective arms, the proportion to which the plan is traced will be varied.

247. A mode of releasing a sounding-weight. When the piece projecting from the bottom of the rod strikes the bottom of the sea, it is forced upward relatively to the rod, and withdraws the catch from under the weight, which drops off and allows the rod to be lifted without it.

248. Union coupling. A is a pipe with a small flange abutting against the pipe, C, with a screwed end; B a nut which holds them together.

249. Ball-and-socket joint, arranged for tubing.

250. Anti-friction bearing. Instead of a shaft revolving in an ordinary bearing it is sometimes supported on the circumference of wheels. The friction is thus reduced to the least amount.

251. Releasing-hook, used in pile-driving machines. When the weight, W, is sufficiently raised, the upper ends of the hooks, A, by which it is suspended, are pressed inward by the sides of the slot, B, in the top of the frame; the weight is thus suddenly released, and falls with accumulating force on to the pile-head.

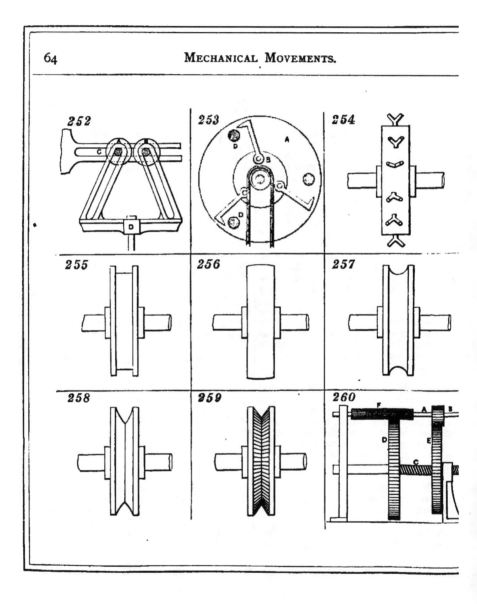

252. A and B are two rollers which require to be equally moved to and fro in the slot, C. This is accomplished by moving the piece, D, with oblique slotted arms, up and down.

253. Centrifugal check-hooks, for preventing accidents in case of the breakage of machinery which raises and lowers workmen, ores, etc., in mines. A is a frame-work fixed to the side of the shaft of the mine, and having fixed studs, D, attached. The drum on which the rope is wound is provided with a flange, B, to which the check-hooks are attached. If the drum acquires a dangerously rapid motion, the hooks fly out by centrifugal force, and one or other or all of them catch hold of the studs, D, and arrest the drum and stop the descent of whatever is attached to the rope. The drum ought besides this to have a spring applied to it, otherwise the jerk arising from the sudden stoppage of the rope might produce worse effects than its rapid motion.

254. A sprocket-wheel to drive or to be driven by a chain.

255. A flanged pulley to drive or be driven by a flat belt.

256. A plain pulley for a flat belt.

257. A concave-grooved pulley for a round band.

258. A smooth-surface V-grooved pulley for a round band.

259. A V-grooved pulley having its groove notched to increase the adhesion of the band.

260. A differential movement. The screw, C, works in a nut secured to the hub of the wheel, E, the nut being free to turn in a bearing in the shorter standard, but prevented by the bearing from any lateral motion. The screw-shaft is secured in the wheel, D. The driving-shaft, A, carries two pinions, F and B. If these pinions were of such size as to turn the two wheels, D and E, with an equal velocity, the screw would remain at rest; but the said wheels being driven at unequal velocities, the screw travels according to the difference of velocity.

261

262

263

264

265

266

267

268

269

261. A combination movement, in which the weight, W, moves vertically with a reciprocating movement; the down-stroke being shorter than the up-stroke. B is a revolving disk, carrying a drum which winds round itself the cord, D. An arm, C, is jointed to the disk and to the upper arm, A, so that when the disk revolves the arm, A, moves up and down, vibrating on the point, G. This arm carries with it the pulley, E. Suppose we detach the cord from the drum and tie it to a fixed point, and then move the arm, A, up and down, the weight, W, will move the same distance, and in addition the movement given to it by the cord, that is to say, the movement will be doubled. Now let us attach the cord to the drum and revolve the disk, B, and the weight will move vertically with the reciprocating motion, in which the down-stroke will be shorter than the up-stroke, because the drum is continually taking up the cord.

262 and 263. The first of these figures is an end view, and the second a side view, of an arrangement of mechanism for obtaining a series of changes of velocity and direction. D is a screw on which is placed eccentrically the cone, B, and C is a friction-roller which is pressed against the cone by a spring or weight. Continuous rotary motion, at a uniform velocity, of the screw, D, carrying the eccentric cone, gives a series of changes of velocity and direction to the roller, C. It will be understood that during every revolution of the cone the roller would press against a different part of the cone, and that it would describe thereon a spiral of the same pitch as the screw, D. The roller, C, would receive a reciprocating motion, the movement in one direction being shorter than that in the other.

264. Two worm-wheels of equal diameter, but one having one tooth more than the other, both in gear with the same worm. Suppose the first wheel has 100 teeth and the second 101, one wheel will gain one revolution over the other during the passage of 100 × 101 teeth of either wheel across the plane of centers, or during 10,100 revolutions of the worm.

265. Variable motion. If the conical drum has a regular circular motion. and the friction-roller is made to traverse lengthwise, a variable rotary motion of the friction-roller will be obtained.

266. The shaft has two screws of different pitches cut on it, one screwing into a fixed bearing, and the other into a bearing free to move to and fro. Rotary motion of the shaft gives rectilinear motion to the movable bearing, a distance equal to the difference of pitches, at each revolution.

267. Friction pulley. When the rim turns in the opposite direction to the arrow, it gives motion to the shaft by means of the pivoted eccentric arms; but when it turns in the direction of the arrow, the arms turn on their pivots and the shaft is at rest. The arms are held to the rim by springs.

268. Circular into reciprocating motion by means of a crank and oscillating rod.

269. Continued rectilinear movement of the frame with mutilated racks gives an alternate rotary motion to the spur-gear.

270

271

272

273

274

275

276

277

278

270. Anti-friction bearing for a pulley.

271. On vibrating the lever to which the two pawls are attached, a nearly continuous rectilinear motion is given to the ratchet-bar.

272. Rotary motion of the beveled disk cam gives a reciprocating rectilinear motion to the rod bearing on its circumference.

273. Rectilinear into rectilinear motion. When the rods, A and B, are brought together, the rods, C and D, are thrust further apart, and *vice versa.*

274. An engine-governor. The rise and fall of the balls, K, are guided by the parabolic curved arms, B, on which the anti-friction wheels, L, run. The rods, F, connecting the wheels, L, with the sleeve move it up and down the spindle, C, D.

275. Rotary motion of the worm gives a rectilinear motion to the rack.

276. Continuous rotary motion of the cam gives a reciprocating rectilinear motion to the bar. The cam is of equal diameter in every direction measured across its center.

277. Col. Colt's invention for obtaining the movement of the cylinder of a revolving fire-arm by the act of cocking the hammer. As the hammer is drawn back to cock it, the dog, a, attached to the tumbler, acts on the ratchet, b, on the back of the cylinder. The dog is held up to the ratchet by a spring, c.

278. C. R. Otis's safety-stop for the platform of a hoisting apparatus. A are the stationary uprights, and B is the upper part of the platform working between them. The rope, a, by which the platform is hoisted, is attached by a pin, b, and spring, c, and the pin is connected by two elbow levers with two pawls, d, which work in ratchets secured to the uprights, A. The weight of the platform and the tension of the rope keep the pawls out of gear from the ratchets in hoisting or lowering the platform, but in case of the breakage of rope the spring, c, presses down the pin, b, and the attached ends of the levers, and so presses the pawls into the ratchets and stops the descent of the platform.

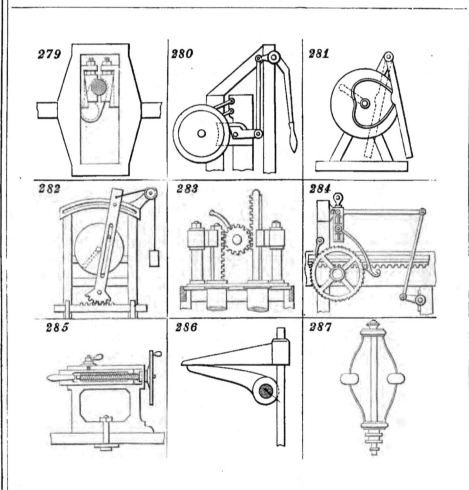

279. Crank and slotted cross-head, with Clayton's sliding journal-box applied to the crank-wrist. This box consists of two taper lining pieces and two taper gibs adjustable by screws, which serve at the same time to tighten the box on the wrist and to set it out to the slot in the cross-head as the box and wrist wear.

280. A mode of working a windlass. By the alternating motion of the long hand-lever to the right, motion is communicated to the short lever, the end of which is in immediate contact with the rim of the wheel. The short lever has a very limited motion upon a pin, which is fixed in a block of cast-iron, which is made with two jaws, each having a flange projecting inward in contact with the inner surface of the rim of the wheel. By the upward motion of the outward end of the short lever, the rim of the wheel is jammed between the end of the lever and the flanges of the block, so as to cause friction sufficient to turn the wheel by the further upward movement of the lever. The backward movement of the wheel is prevented by a common ratchet-wheel and pawls ; as the short lever is pushed down it frees the wheel and slides freely over it.

281. The revolution of the disk causes the lever at the right to vibrate by the pin moving in the groove in the face of the disk.

282. By the revolution of the disk in which is fixed a pin working in a slot in the upright bar which turns on a center near the bottom, both ends of the bar are made to traverse, the toothed sector producing alternate rectilinear motion in the horizontal bar at the bottom, and also alternate perpendicular motion of the weight.

283. By a vibratory motion of the handle, motion is communicated by the pinion to the racks. This is used in working small air pumps for scientific experiments.

284. Represents a feeding apparatus for the bed of a sawing machine. By the revolution of the crank at the lower part of the figure, alternate motion is communicated to the horizontal arm of the bell-crank lever whose fulcrum is at a, near the top left-hand corner of the figure. By this means motion is communicated to the catch attached to the vertical arm of the lever, and the said catch communicates motion to the ratchet-wheel, upon the shaft of which is a toothed pinion, working in the rack attached to the side of the carriage. The feed is varied by a screw in the bell-crank lever.

285. Is the movable head of a turning lathe. By turning the wheel to the right, motion is communicated to the screw, producing rectilinear motion of the spindle in the end of which the center is fixed.

286. Toe and lifter for working puppet valves in steam engines. The curved toe on the rock-shaft operates on the lifter attached to the lifting-rod to raise the valve.

287. Pickering's governor. The balls are attached to springs the upper end of each of which is attached to a collar fixed on the spindle, and the lower end to a collar on the sliding sleeve. The springs yield in a proper degree to the centrifugal force of the balls, and raise the sleeve ; and as the centrifugal force diminishes, they draw the balls toward the spindle and depress the sleeve.

288

289

290

291

292

293

294

295

296

288 and 289. The former is what is termed a *recoil*, and the latter a *repose* or *dead-beat* escapement for clocks. The same letters of reference indicate like parts in both. The *anchor*, H, L, K, is caused, by the oscillation of the pendulum, to vibrate upon the axis, *a*. Between the two extremities, or pallets, H, K, is placed the escape-wheel, A, the teeth of which come alternately against the outer surface of the pallet, K, and inner surface of pallet, H. In 289 these surfaces are cut to a curve concentric to the axis, *a ;* consequently, during the time one of the teeth is against the pallet the wheel remains perfectly at rest. Hence the name *repose* or *dead-beat*. In 288 the surfaces are of a different form, not necessary to explain, as it can be understood that any form not concentric with the axis, *a*, must produce a slight recoil of the wheel during the escape of the tooth, and hence the term *recoil* escapement. On the pallets leaving teeth, at each oscillation of the pendulum, the extremities of teeth slide along the surfaces, *c, e*, and *d, b*, and give sufficient impulse to pendulum.

290. Another kind of pendulum escapement.

291. Arnold's chronometer or free escapement, sometimes used in watches. A spring, A, is fixed or screwed against the plate of the watch at *b*. To the under side of this spring is attached a small stop, *d*, against which rest successively the teeth of the escape-wheel, B ; and on the top of spring is fixed a stud, *i*, holding a lighter and more flexible spring which passes under a hook, *k*, at the extremity of A, so that it is free on being depressed, but in rising would lift A. On the axis of the balance is a small stud, *a*, which touches the thin spring at each oscillation of balance-wheel. When the movement is in the direction shown by the arrow, the stud depresses the spring in passing, but on returning raises it and the spring, A, and stop, *d*, and thus allows one tooth of escape-wheel to pass, letting them fall immediately to arrest the next. At the same time that this tooth escapes another strikes against the side of the notch, *g*, and restores to balance-wheel the force lost during a vibration. It will be understood that only at one point is the free movement of balance opposed during an oscillation.

292. Stud escapement, used in large clocks. One pallet, B, works in front of the wheel and the other at the back. The studs are arranged in the same manner, and rest alternately upon the front or back pallet. As the curve of the pallets is an arc described from F, this is a *repose* or *dead-beat* escapement.

293. Duplex escapement, for watches, so called from partaking of the characters of the spur and crown wheels. The axis of balance carries pallet, B, which at every oscillation receives an impulse from the crown teeth. In the axis, A, of balance-wheel is cut a notch into which the teeth round the edge of the wheel successively fall after each one of the crown teeth passes the impulse pallet, B.

294 and 295. A cylinder escapement. 294 shows the cylinder in perspective, and 295 shows part of the escape-wheel on a large scale, and represents the different positions taken by cylinder, A, B, during an oscillation. The pallets, *a, b, c*, on the wheel rest alternately on the inside and outside of cylinder. To the top of cylinder is attached the balance-wheel. The wheel pallets are beveled so as to keep up the impulse of balance by sliding against the beveled edge of cylinder.

296. Lever escapement. The anchor or piece, B, which carries the pallets, is attached to lever, E, C, at one end of which is a notch, E. On a disk secured on the arbor of balance is fixed a small pin which enters the notch at the middle of each vibration, causing the pallet to enter in and retire from between the teeth of escape-wheel The wheel gives an impulse to each of the pallets alternately as it leaves a tooth, and the lever gives impulse to the balance-wheel in opposite directions alternately.

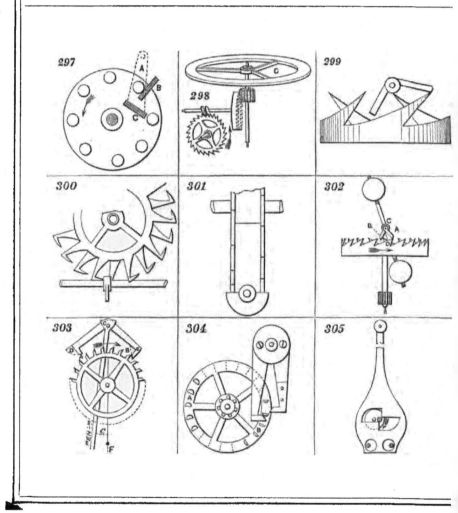

297. An escapement with a lantern wheel. An arm, A, carries the two pallets, B and C.

298. An old-fashioned watch escapement.

299. An old-fashioned clock escapement.

300 and 301. A clock or watch escapement; 300 being a front elevation, and 301 a side elevation. The pallet is acted upon by the teeth of one and the other of two escape-wheels alternately.

302. Balance-wheel escapement. C is the balance; A, B, are the pallets; and D is the escape-wheel.

303. A dead-beat pendulum escapement. The inner face of the pallet, E, and outer face of D, are concentric with the axis on which the pallets vibrate, and hence there is no recoil.

304. Pin-wheel escapement, somewhat resembling the stud escapement shown by 292. The pins, A, B, of the escape-wheel are of two different forms, but the form of those on the right side is the best. One advantage of this kind of escapement is that if one of the pins is damaged it can easily be replaced, whereas if a tooth is damaged the whole wheel is ruined.

305. A single-pin pendulum escapement. The escape-wheel is a very small disk with single eccentric pin; it makes half a revolution for every beat of the pendulum, giving the impulse on the upright faces of the pallets, the horizontal faces of which are dead ones. This can also be adapted to watches.

306

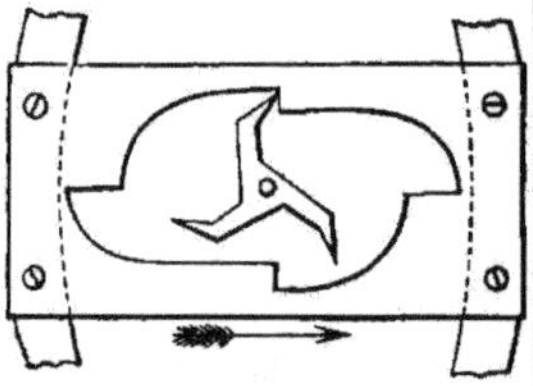

307

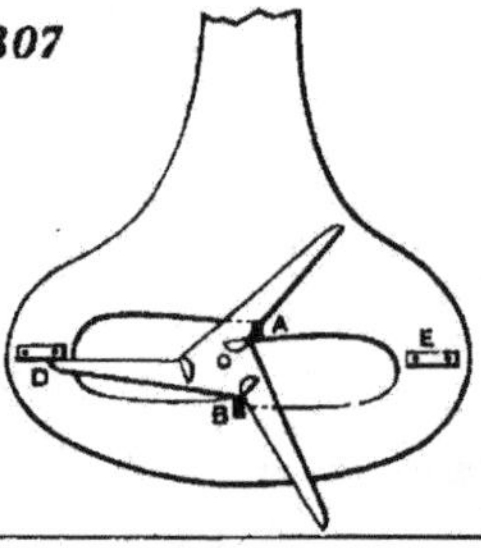

308

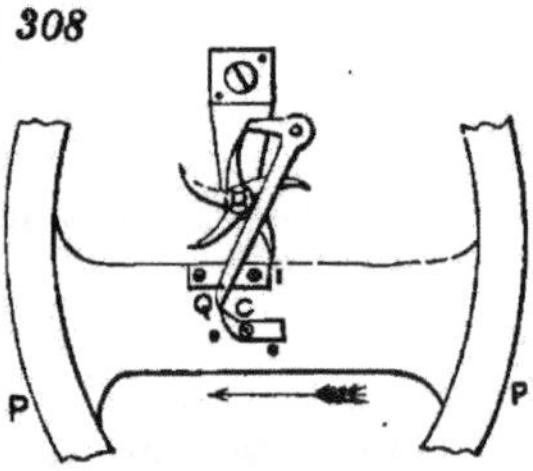

P　　　　　　　　P

309

310

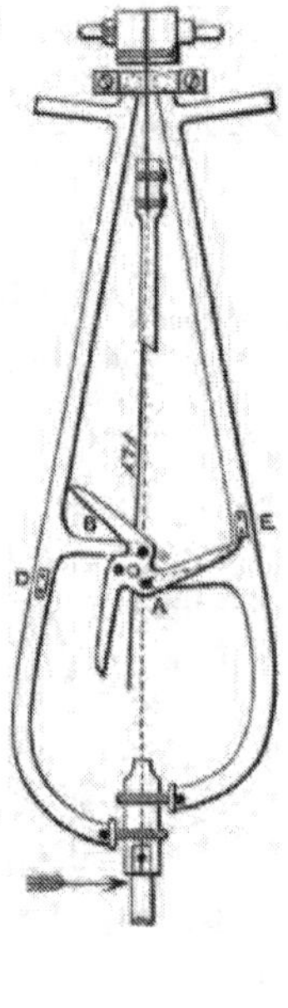

311

FLY

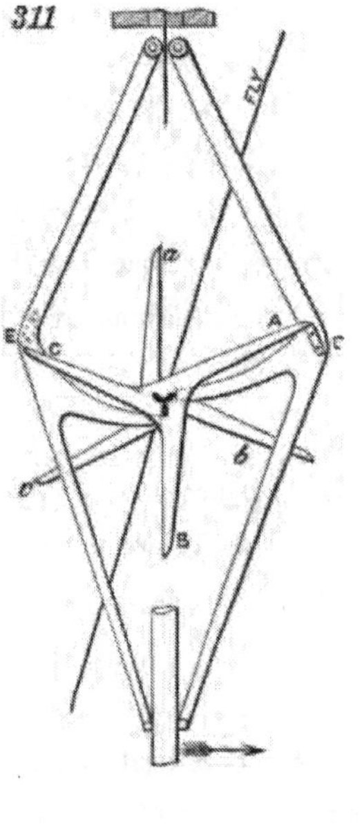

312

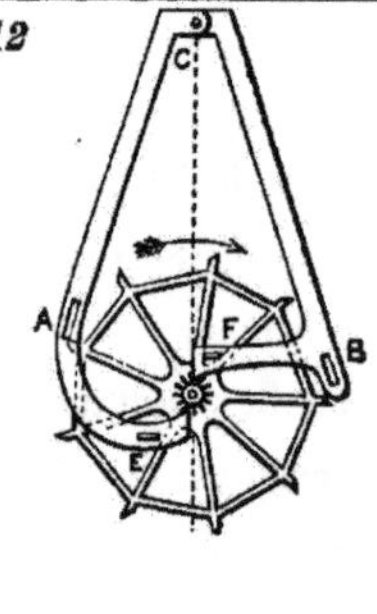

306. Three-legged pendulum escapement. The pallets are formed in an opening in a plate attached to the pendulum, and the three teeth of the escape-wheel operate on the upper and lower pallets alternately. One tooth is shown in operation on the upper pallet.

307. A modification of the above with long stopping teeth, D and E. A and B are the pallets.

308. A detached pendulum escapement, leaving the pendulum, P, free or detached from the escape-wheel, except at the time of receiving the impulse and unlocking the wheel. There is but one pallet, I, which receives impulse only during the vibrations of the pendulum to the left. The lever, Q, locks the escape-wheel until just before the time for giving the impulse, when it is unlocked by the click, C, attached to the pendulum. As the pendulum returns to the right, the click, which oscillates on a pivot, will be pushed aside by the lever.

309. Mudge's gravity escapement. The pallets, A, B, instead of being on one arbor, are on two, as shown at C. The pendulum plays between the fork-pins, P, Q, and so raises one of the weighted pallets out of the wheel at each vibration. When the pendulum returns the pallet falls with it, and the weight of the pallet gives the impulse.

310. Three-legged gravity escapement. The lifting of the pallets, A and B, is done by the three pins near the center of the escape-wheel, the pallets vibrating from two centers near the point of suspension of the pendulum. The escape-wheel is locked by means of stops, D and E, on the pallets.

311. Double three-legged gravity escapement. Two locking-wheels, A, B, C, and a, b, c, are here used with one set of lifting-pins between them. The two wheels are set wide enough apart to allow the pallets to lie between them. The teeth of the first-mentioned locking-wheel are stopped by a stop-tooth, D, on one pallet, and those of the other one by a stop-tooth, E, on the other pallet.

312. Bloxam's gravity escapement. The pallets are lifted alternately by the small wheel, and the stopping is done by the action of the stops, A and B, on the larger wheel. E and F are the fork-pins which embrace the pendulum.

313

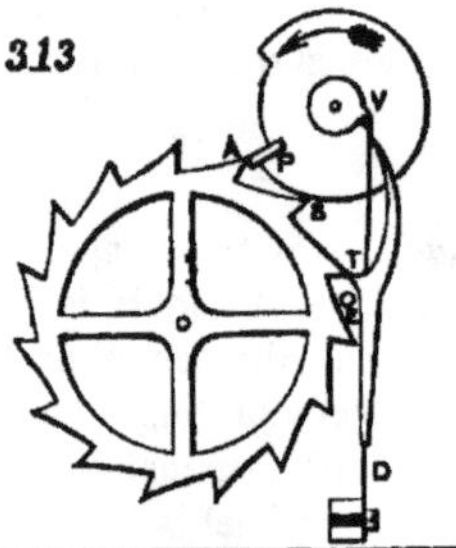

314

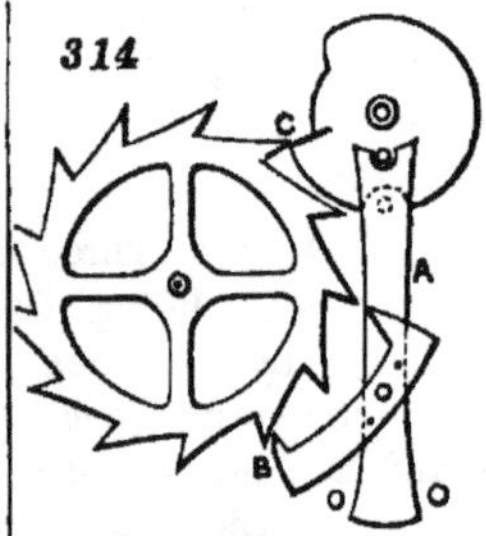

315

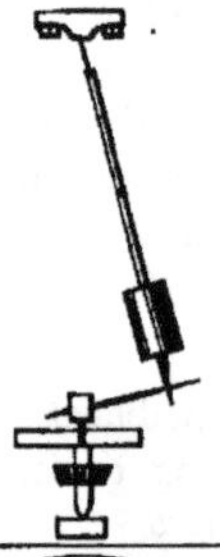

316

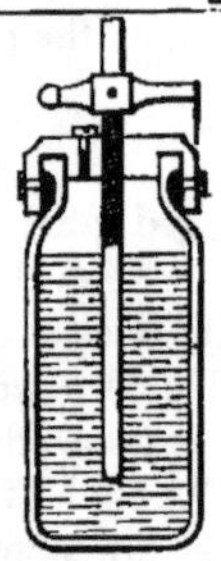

317

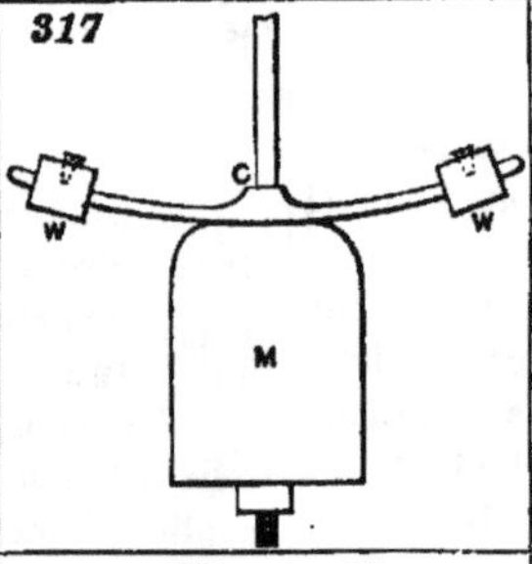

318

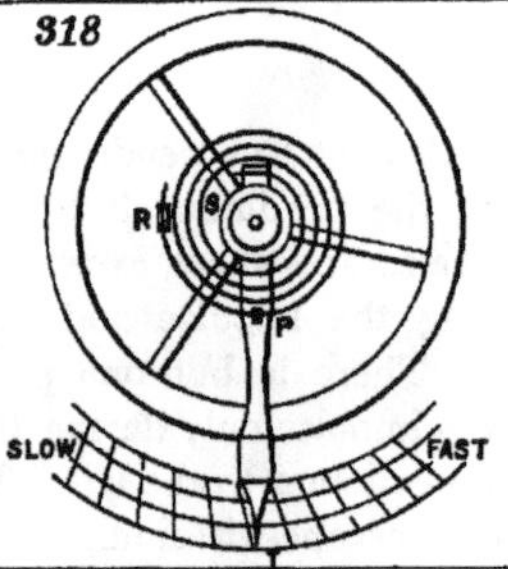

319

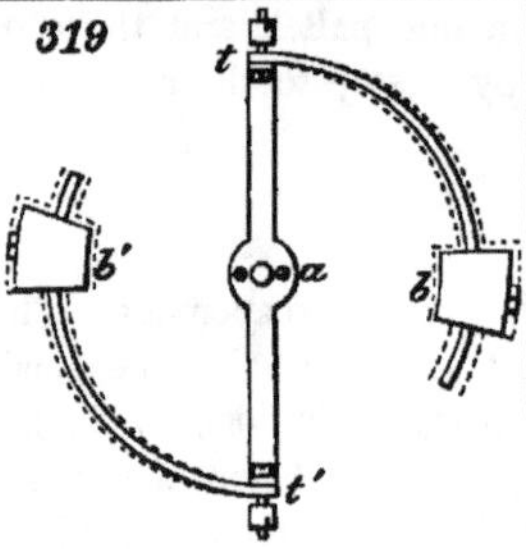

320

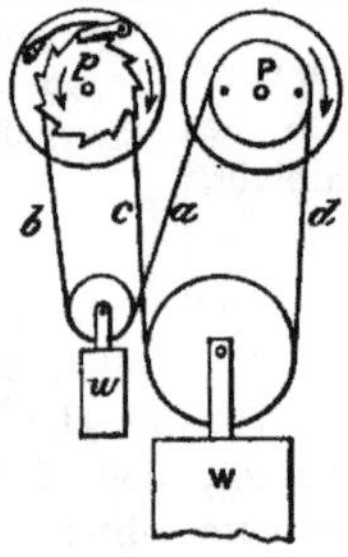

321

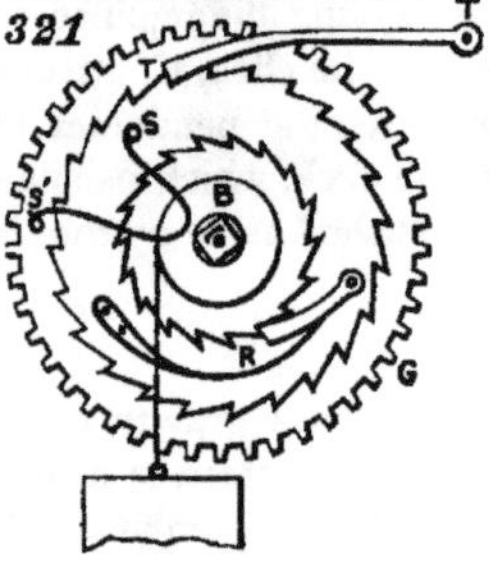

313. Chronometer escapement, the form now commonly constructed. As the balance rotates in the direction of the arrow, the tooth, V, on the verge, presses the passing-spring against the lever, pressing aside the lever and removing the detent from the tooth of the escape-wheel. As balance returns, tooth, V, presses aside and passes spring without moving lever, which then rests against the stop, E. P is the only pallet upon which impulse is given.

314. Lever chronometer escapement. In this the pallets, A, B, and lever, look like those of the lever escapement 296: but these pallets only lock the escape-wheel, having no impulse. Impulse is given by teeth of escape-wheel directly to a pallet, C, attached to balance.

315. Conical pendulum, hung by a thin piece of round wire. Lower end connected with and driven in a circle by an arm attached to a vertical rotating spindle. The pendulum-rod describes a cone in its revolution.

316. Mercurial compensation pendulum. A glass jar of mercury is used for the bob or weight. As the pendulum-rod is expanded lengthwise by increased temperature, the expansion of mercury in jar carries it to a greater height therein, and so raises its center of gravity relatively to the rod sufficiently to compensate for downward expansion of the rod. As rod is contracted by a reduction of temperature, contraction of mercury lowers it relatively to rod. In this way the center of oscillation is always kept in the same place, and the effective length of pendulum always the same.

317. Compound bar compensation pendulum. C is a compound bar of brass and iron or steel, brazed together with brass downward. As brass expands more than iron, the bar will bend upward as it gets warmer, and carry the weights, W, W, up with it, raising the center of the aggregate weight, M, W, to raise the center of oscillation as much as elongation of the pendulum-rod would let it down.

318. Watch regulator. The balance-spring is attached at its outer end to a fixed stud, R, and at its inner end to staff of balance. A neutral point is formed in the spring at P by inserting it between two curb-pins in the lever, which is fitted to turn on a fixed ring concentric with staff of balance, and the spring only vibrates between this neutral point and staff of balance. By moving lever to the right, the curb-pins are made to reduce the length of acting part of spring, and the vibrations of balance are made faster; and by moving it to the left an opposite effect is produced.

319. Compensation balance. *t, a, t'*, is the main bar of balance, with timing screws for regulation at the ends. *t* and *t'* are two compound bars, of which the outside is brass and the inside steel, carrying weights, *b, b'*. As heat increases, these bars are bent inward by the greater expansion of the brass, and the weights are thus drawn inward, diminishing the inertia of the balance. As the heat diminishes, an opposite effect is produced. This balance compensates both for its own expansion and contraction, and that of the balance-spring.

320. Endless chain, maintaining power on going-barrel, to keep a clock going while winding, during which operation the action of the weight or main-spring is taken off the barrel. The wheel to the right is the "going-wheel," and that to the left the "striking-wheel." P is a pulley fixed to the great wheel of the going part, and roughened, to prevent a rope or chain hung over it from slipping. A similar pulley rides on another arbor, *p*, which may be the arbor of the great wheel of the striking part, and attached by a ratchet and click to that wheel, or to clock-frame, if there is no striking part. The weights are hung, as may be seen, the small one being only large enough to keep the rope or chain on the pulleys. If the part, *b*, of the rope or chain is pulled down, the ratchet-pulley runs under the click, and the great weight is pulled up by *c*, without taking its pressure off the going-wheel at all.

321. Harrison's "going-barrel." Larger ratchet-wheel, to which the click, R, is attached, is connected with the great wheel, G, by a spring, S, S'. While the clock is going, the weight acts upon the great wheel, G, through the spring; but as soon as the weight is taken off by winding, the click, T, whose pivot is set in the frame, prevents the larger ratchet from falling back, and so the spring, S, S', still drives the great wheel during the time the clock takes to wind, as it need only just keep the escapement going, the pendulum taking care of itself for that short time. Good watches have a substantially similar apparatus.

322

323

324

325

326

327

328

329

330

322. A very convenient construction of parallel ruler for drawing, made by cutting a quadrangle through the diagonal, forming two right-angled triangles, A and B. It is used by sliding the hypothenuse of one triangle upon that of the other.

323. Parallel ruler consisting of a simple straight ruler, B, with an attached axle, C, and pair of wheels, A, A. The wheels, which protrude but slightly through the under side of the ruler, have their edges nicked to take hold of the paper and keep the ruler always parallel with any lines drawn upon it.

324. Compound parallel ruler, composed of two simple rulers, A, A, connected by two crossed arms pivoted together at the middle of their length, each pivoted at one end to one of the rulers, and connected with the other one by a slot and sliding-pin, as shown at B. In this the ends as well as the edges are kept parallel. The principle of construction of the several rulers represented is taken advantage of in the formation of some parts of machinery.

325. Parallel ruler composed of two simple rulers, A, B, connected by two pivoted swinging arms, C, C.

326. A simple means of guiding or obtaining a parallel motion of the piston-rod of an engine. The slide, A, moves in and is guided by the vertical slot in the frame, which is planed to a true surface.

327. Differs from 326 in having rollers substituted for the slides on the cross-head, said rollers working against straight guide-bars, A, A, attached to the frame. This is used for small engines in France.

328. A parallel motion invented by Dr. Cartwright in the year 1787. The toothed wheels, C, C, have equal diameters and numbers of teeth; and the cranks, A, A, have equal radii, and are set in opposite directions, and consequently give an equal obliquity to the connecting-rods during the revolution of the wheels. The cross-head on the piston-rod being attached to the two connecting-rods, the piston-rod is caused to move in a right line.

329. A piston-rod guide. The piston-rod, A, is connected with a wrist attached to a cog-wheel, B, which turns on a crank-pin, carried by a plate, C, which is fast on the shaft. The wheel, B, revolves around a stationary internally toothed gear, D, of double the diameter of B, and so motion is given to the crank-pin, and the piston-rod is kept upright.

330. The piston-rod is prolonged and works in a guide, A, which is in line with the center of the cylinder. The lower part of the connecting-rod is forked to permit the upper part of the piston-rod to pass between.

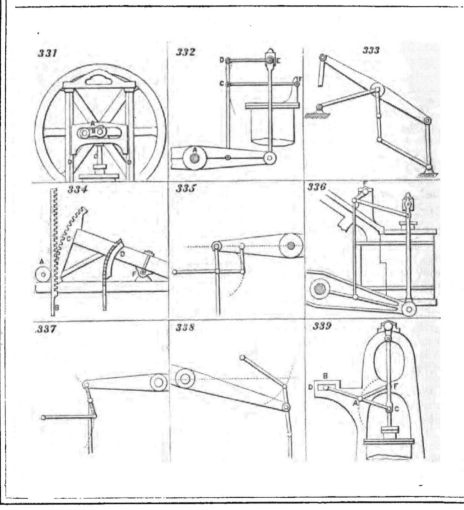

331. An engine with crank motion like that represented in 93 and 279 of this table, the crank-wrist journal working in a slotted cross-head, A. This cross-head works between the pillar guides, D, D, of the engine framing.

332. A parallel motion used for the piston-rod of side lever marine engines. F, C, is the radius bar, and E the cross-head to which the parallel bar, E, D, is attached.

333. A parallel motion used only in particular cases.

334. Shows a parallel motion used in some of the old single-acting beam engines. The piston-rod is formed with a straight rack gearing with a toothed segment on the beam. The back of the rack works against a roller, A.

335. A parallel motion commonly used for stationary beam engines.

336. An arrangement of parallel motion for side lever marine engines. The parallel rods connected with the side rods from the beams or side levers are also connected with short radius arms on a rock-shaft working in fixed bearings.

337. Parallel motion in which the radius rod is connected with the lower end of a short vibrating rod, the upper end of which is connected with the beam, and to the center of which the piston-rod is connected.

338. Another modification, in which the radius bar is placed above the beam.

339. Parallel motion for direct action engines. In this, the end of the bar, B, C, is connected with the piston-rod, and the end, B, slides in a fixed slot, D. The radius bar, F, A, is connected at F with a fixed pivot, and at A, midway between the ends of B, C.

340

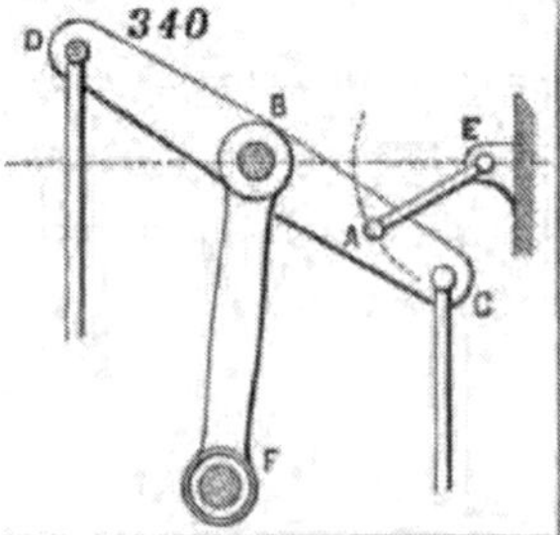

341

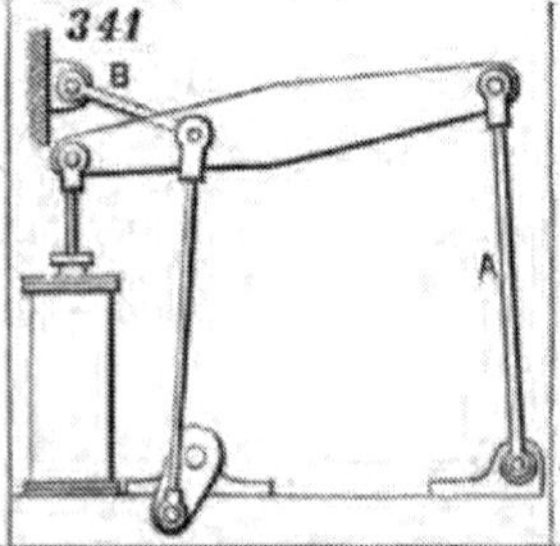

342

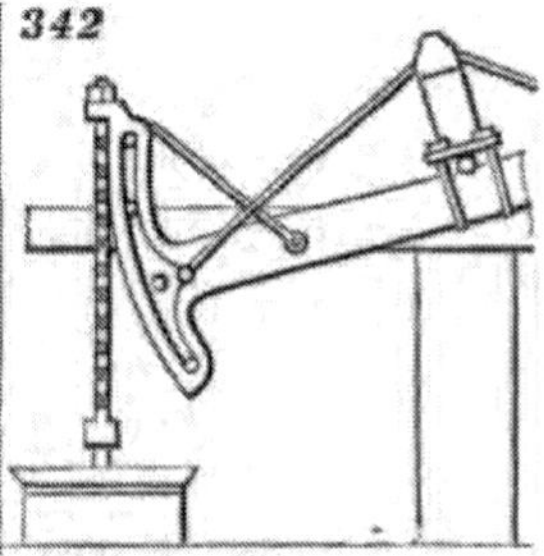

343

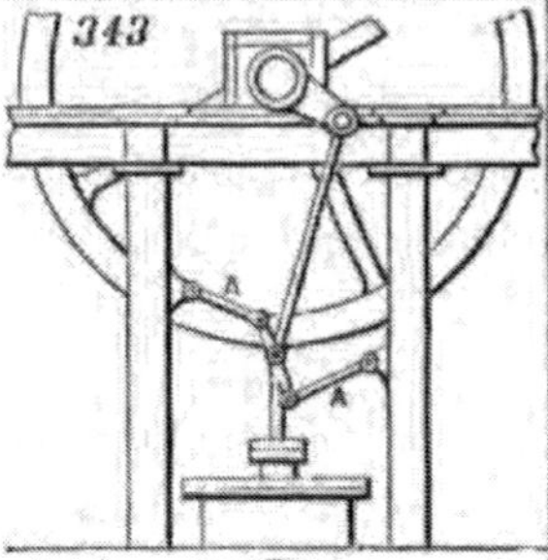

344

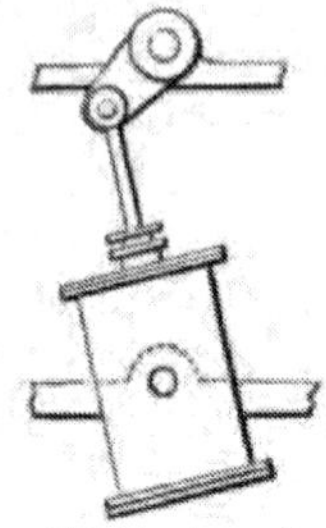

345

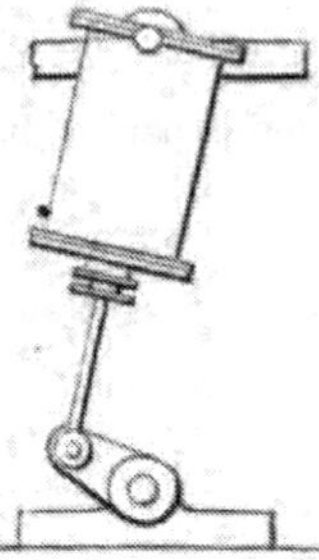

346

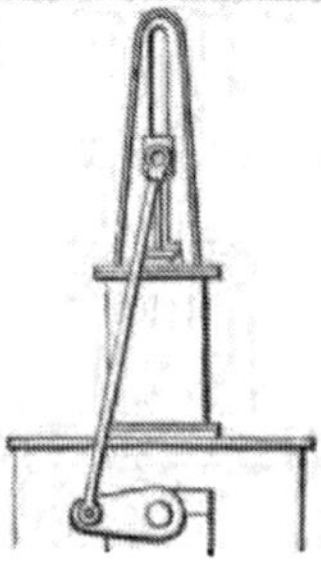

347

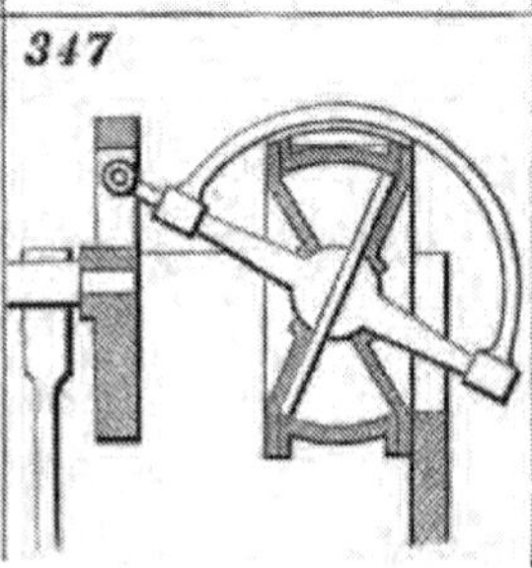

348

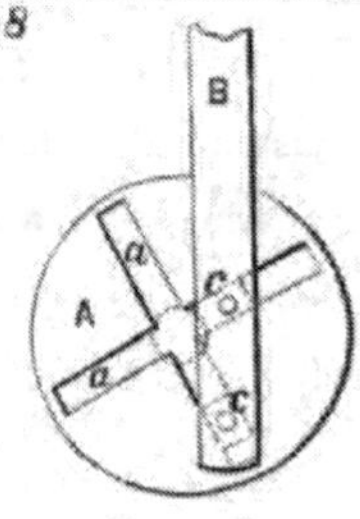

340. Another parallel motion. Beam, D, C, with joggling pillar-support, B, F, which vibrates from the center, F. The piston-rod is connected at C. The radius-bar, E, A, produces the parallel motion.

341. "Grasshopper" beam engine. The beam is attached at one end to a rocking-pillar, A, and the shaft arranged as near to the cylinder as the crank will work. B is the radius-bar of the parallel motion.

342. Old-fashioned single-acting beam pumping engine on the atmospheric principle, with chain connection between piston-rod and a segment at end of beam. The cylinder is open at top. Very low pressure steam is admitted below piston, and the weight of pump-rod, etc., at the other end of beam, helps to raise piston. Steam is then condensed by injection, and a vacuum thus produced below piston, which is then forced down by atmospheric pressure thereby drawing up pump-rod.

343. Parallel motion for upright engine. A, A, are radius-rods connected at one end with the framing and at the other with a vibrating piece on top of piston-rod.

344. Oscillating engine. The cylinder has trunnions at the middle of its length working in fixed bearings, and the piston-rod is connected directly with the crank, and no guides are used.

345. Inverted oscillating or pendulum engine. The cylinder has trunnions at its upper end and swings like a pendulum. The crank-shaft is below, and the piston-rod connected directly with crank.

346. Table engine. The cylinder is fixed on a table-like base. The piston-rod has a cross-head working in straight slotted guides fixed on top of cylinder, and is connected by two side connecting-rods with two parallel cranks on shaft under the table.

347. Section of disk engine. Disk piston, seen edgewise, has a motion substantially like a coin when it first falls after being spun in the air. The cylinder-heads are cones. The piston-rod is made with a ball to which the disk is attached, said ball working in concentric seats in cylinder-heads, and the left-hand end is attached to the crank-arm or fly-wheel on end of shaft at left. Steam is admitted alternately on either side of piston.

348. Mode of obtaining two reciprocating movements of a rod by one revolution of a shaft, patented in 1836 by B. F. Snyder, has been used for operating the needle of a sewing machine, by J. S. McCurdy, also for driving a gang of saws. The disk, A, on the central rotating shaft has two slots, a, a, crossing each other at a right angle in the center, and the connecting-rod, B, has attached to it two pivoted slides, c, c, one working in each slot.

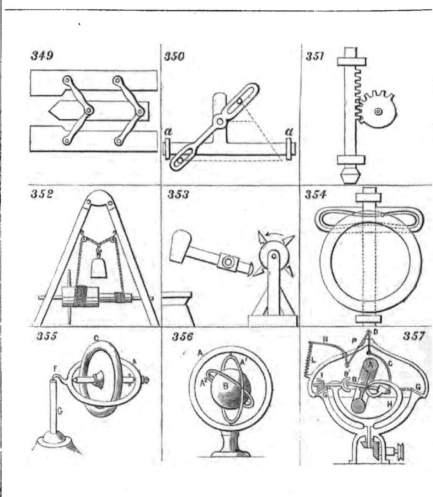

349. Another form of parallel ruler. The arms are jointed in the middle and connected with an intermediate bar, by which means the ends of the ruler, as well as the sides, are kept parallel.

350. Traverse or to-and-fro motion. The pin in the upper slot being stationary, and the one in the lower slot made to move in the direction of the horizontal dotted line, the lever will by its connection with the bar give to the latter a traversing motion in its guides, a, a.

351. Stamp. Vertical percussive falls derived from horizontal rotating shaft. The mutilated toothed pinion acts upon the rack to raise the rod until its teeth leave the rack and allow the rod to fall.

352. Another arrangement of the Chinese windlass illustrated by 129 of this table.

353. A modification of the tilt or trip hammer, illustrated by 74. In this the hammer helve is a lever of the first order. In 74 it is a lever of the third order.

354. A modification of the crank and slotted cross-head, 93. The cross-head contains an endless groove in which the crank-wrist works, and which is formed to produce a uniform velocity of movement of the wrist or reciprocating-rod.

355. The gyroscope or rotascope, an instrument illustrating the tendency of rotating bodies to preserve their plane of rotation. The spindle of the metallic disk, C, is fitted to turn easily in bearings in the ring, A. If the disk is set in rapid rotary motion on its axis, and the pintle, F, at one side of the ring, A, is placed on the bearing in the top of the pillar, G, the disk and ring seem indifferent to gravity, and instead of dropping begin to revolve about the vertical axis.

356. Bohnenberger's machine illustrating the same tendency of rotating bodies. This consists of three rings, A, A^1, A^2, placed one within the other and connected by pivots at right angles to each other. The smallest ring, A^2, contains the bearings for the axis of a heavy ball, B. The ball being set in rapid rotation, its axis will continue in the same direction, no matter how the position of the rings may be altered; and the ring, A^2, which supports it will resist a considerable pressure tending to displace it.

357. What is called the gyroscope governor, for steam engines, etc., patented by Alban Anderson in 1858. A is a heavy wheel, the axle, B, B^1, of which is made in two pieces connected together by a universal joint. The wheel, A, is on one piece, B, and a pinion, I, on the other piece, B^1. The piece, B, is connected at its middle by a hinge joint with the revolving frame, H, so that variations in the inclination of the wheel, A, will cause the outer end of the piece, B, to rise and fall. The frame, H, is driven by bevel gearing from the engine, and by that means the pinion, I, is carried round the stationary toothed circle, G, and the wheel, A, is thus made to receive a rapid rotary motion on its axis. When the frame, H, and wheel, A, are in motion, the tendency of the wheel, A, is to assume a vertical position, but this tendency is opposed by a spring, L. The greater the velocity of the governor, the stronger is the tendency above mentioned, and the more it overcomes the force of the spring, and *vice versa*. The piece, B, is connected with the valve-rod by rods, C, D, and the spring, L, is connected with the said rod by levers, N, and rod, P.

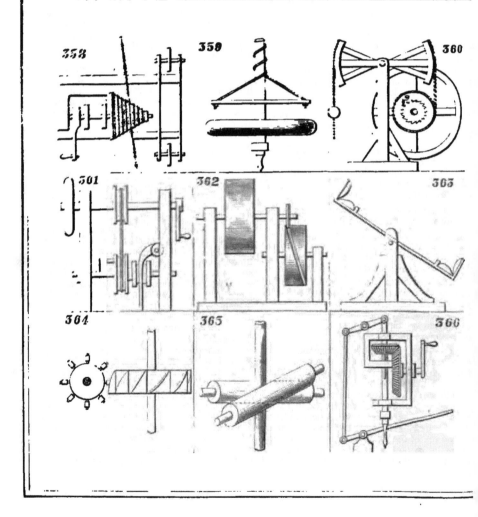

358. Traverse of carriage, made variable by fusee according to the variation in diameter where the band acts.

359. Primitive drilling apparatus. Being once set in motion, it is kept going by hand, by alternately pressing down and relieving the transverse bar to which the bands are attached, causing the bands to wind upon the spindle alternately in opposite directions, while the heavy disk or fly-wheel gives a steady momentum to the drill-spindle in its rotary motion.

360. Continuous rotary motion from oscillating. The beam being made to vibrate, the drum to which the cord is attached, working loose on fly-wheel shaft, gives motion to said shaft through the pawl and ratchet-wheel, the pawl being attached to drum and the ratchet-wheel fast on shaft.

361. Another simple form of clutch for pulleys, consisting of a pin on the lower shaft and a pin on side of pulley. The pulley is moved lengthwise of the shaft by means of a lever or other means to bring its pin into or out of contact with the pin on shaft.

362. Alternating traverse of upper shaft and its drum, produced by pin on the end of the shaft working in oblique groove in the lower cylinder.

363. See-saw, one of the simplest illustrations of a limited oscillating or alternate circular motion.

364. Intermittent rotary motion from continuous rotary motion about an axis at right angles. Small wheel on left is driver ; and the friction rollers on its radial studs work against the faces of oblique grooves or projections across the face of the larger wheel, and impart motion thereto.

365. Cylindrical rod arranged between two rollers, the axes of which are oblique to each other. The rotation of the rollers produces both a longitudinal and a rotary motion of the rod.

366. Drilling machine. By the large bevel-gear rotary motion is given to vertical drill-shaft, which slides through small bevel-gear but is made to turn with it by a feather and groove, and is depressed by treadle connected with upper lever.

367

368

369

370

371

372

373

374

375

367. A parallel ruler with which lines may be drawn at required distances apart without setting out. Lower edge of upper blade has a graduated ivory scale, on which the incidence of the outer edge of the brass arc indicates the width between blades.

368. Describing spiral line on a cylinder. The spur-gear which drives the bevel-gears, and thus gives rotary motion to the cylinder, also gears into the toothed rack, and thereby causes the marking point to traverse from end to end of the cylinder.

369. Cycloidal surfaces, causing pendulum to move in cycloidal curve, rendering oscillations isochronous or equal-timed.

370. Motion for polishing mirrors, the rubbing of which should be varied as much as practicable. The handle turns the crank to which the long bar and attached ratchet-wheel are connected. The mirror is secured rigidly to the ratchet-wheel. The long bar, which is guided by pins in the lower rail, has both a longitudinal and an oscillating movement, and the ratchet-wheel is caused to rotate intermittently by a click operated by an eccentric on the crank-shaft, and hence the mirror has a compound movement.

371. Modification of mangle-wheel motion. The large wheel is toothed on both faces, and an alternating circular motion is produced by the uniform revolution of the pinion, which passes from one side of the wheel to the other through an opening on the left of the figure.

372. White's dynamometer, for determining the amount of power required to give rotary motion to any piece of mechanism. The two horizontal bevel-gears are arranged in a hoop-shaped frame, which revolves freely on the middle of the horizontal shaft, on which there are two vertical bevel-gears gearing to the horizontal ones, one fast and the other loose on the shaft. Suppose the hoop to be held stationary, motion given to either vertical bevel-gear will be imparted through the horizontal gears to the other vertical one; but if the hoop be permitted it will revolve with the vertical gear put in motion, and the amount of power required to hold it stationary will correspond with that transmitted from the first gear, and a band attached to its periphery will indicate that power by the weight required to keep it still.

373. Robert's contrivance for proving that friction of a wheel carriage does not increase with velocity, but only with load. Loaded wagon is supported on surface of large wheel, and connected with indicator constructed with spiral spring, to show force required to keep carriage stationary when large wheel is put in motion. It was found that difference in velocity produced no variation in the indicator, but difference in weight immediately did so.

374. Rotary motion of shaft from treadle by means of an endless band running from a roller on the treadle to an eccentric on the shaft.

375. Pair of edge runners or chasers for crushing or grinding. The axles are connected with vertical shaft, and the wheels or chasers run in an annular pan or trough.

376

377

378

379

380

381

382

383

384

376. Tread-wheel horse-power turned by the weight of an animal attempting to walk up one side of its interior; has been used for driving the paddle-wheels of ferry-boats and other purposes by horses. The turn-spit dog used also to be employed in such a wheel in ancient times for turning meat while roasting on a spit.

377. The tread-mill employed in jails in some countries for exercising criminals condemned to labor, and employed in grinding grain, etc.; turns by weight of persons stepping on tread-boards on periphery. This is supposed to be a Chinese invention, and it is still used in China for raising water for irrigation.

378. Saw for cutting trees by motion of pendulum, is represented as cutting a lying tree.

379 and 380. Portable cramp drills. In 379 the feed-screw is opposite the drill, and in 380 the drill spindle passes through the center of the feed-screw.

381. Bowery's joiner's clamp, plan and transverse section. Oblong bed has, at one end, two wedge-formed cheeks, adjacent sides of which lie at an angle to each other, and are dovetailed inward from upper edge to receive two wedges for clamping the piece or pieces of wood to be planed.

382. Adjustable stand for mirrors, etc., by which a glass or other article can be raised or lowered, turned to the right or left, and varied in its inclination. The stem is fitted into a socket of pillar, and secured by a set screw, and the glass is hinged to the stem, and a set screw is applied to the hinge to tighten it. The same thing is used for photographic camera-stands.

383. Represents the principal elements of machinery for dressing cloth and warps, consisting of two rollers, from one to the other of which the yarn or cloth is wound, and an interposed cylinder having its periphery either smooth-surfaced or armed with brushes, teasels, or other contrivances, according to the nature of the work to be done. These elements are used in machines for sizing warps, gig-mills for dressing woolen goods, and in most machines for finishing woven fabrics.

384. Helicograph, or instrument for describing helices. The small wheel, by revolving about the fixed central point, describes a volute or spiral by moving along the screw-threaded axle either way, and transmits the same to drawing paper on which transfer-paper is laid with colored side downward.

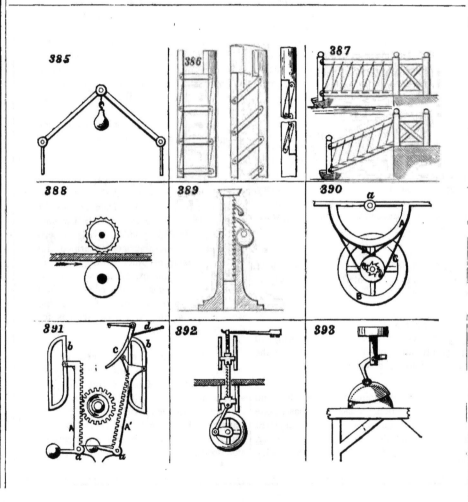

385. Contrivance employed in Russia for shutting doors. One pin is fitted to and turns in socket attached to door, and the other is similarly attached to frame. In opening the door, pins are brought together, and weight is raised. Weight closes door by depressing the joint of the toggle toward a straight line, and so widening the space between the pins.

386. Folding library ladder. It is shown open, partly open, and closed; the rounds are pivoted to the side-pieces, which are fitted together to form a round pole when closed, the rounds shutting up inside.

387. Self-adjusting step-ladder for wharfs at which there are rise and fall of tide. The steps are pivoted at one edge into wooden bars forming string-pieces, and their other edge is supported by rods suspended from bars forming hand-rails. The steps remain horizontal whatever position the ladder assumes.

388. Feed-motion of Woodworth's planing machine, a smooth supporting roller, and a toothed top roller.

389. Lifting-jack operated by an eccentric, pawl, and ratchet. The upper pawl is a stop.

390. Device for converting oscillating into rotary motion. The semicircular piece, A, is attached to a lever which works on a fulcrum, a, and it has attached to it the ends of two bands, C and D, which run around two pulleys, loose on the shaft of the fly-wheel, B. Band, C, is open, and band, D, crossed. The pulleys have attached to them pawls which engage with two ratchet-wheels fast on the fly-wheel shaft. One pawl acts on its ratchet-wheel when the piece, A, turns one way, and the other when the said piece turns the other way, and thus a continuous rotary motion of the shaft is obtained.

391. Reciprocating into rotary motion. The weighted racks, A, A[1], are pivoted to the end of a piston-rod, and pins at the end of the said racks work in fixed guide-grooves, b, b, in such manner that one rack operates upon the cog-wheel in ascending and the other in descending, and so continuous rotary motion is produced. The elbow lever, C, and spring, d, are for carrying the pin of the right-hand rack over the upper angle in its guide-groove, b.

392. Gig-saw, the lower end connected with a crank which works it, and the upper end connected with a spring which keeps it strained without a gate.

393. Contrivance for polishing lenses and bodies of spherical form. The polishing material is in a cup connected by a ball-and-socket joint and bent piece of metal with a rotating upright shaft set concentric to the body to be polished. The cup is set eccentric, and by that means is caused to have an independent rotary motion about its axis on the universal joint, as well as to revolve about the common axis of the shaft and the body to be polished. This prevents the parts of the surface of the cup from coming repeatedly in contact with the same parts of surface of the lens or other body

394

395

396

397

398

399

400

401

402

394. C. Parsons's patent device for converting reciprocating motion into rotary, an endless rack provided with grooves on its side gearing with a pinion having two concentric flanges of different diameters. A substitute for crank in oscillating cylinder engines.

395. Four-way cock, used many years ago on steam engines to admit and exhaust steam from the cylinder. The two positions represented are produced by a quarter turn of the plug. Supposing the steam to enter at the top, in the upper figure the exhaust is from the right end of the cylinder, and in the lower figure the exhaust is from the left—the steam entering, of course, in the opposite port.

396. G. P. Reed's patent anchor and lever escapement for watches. The lever is so applied in combination with chronometer escapement that the whole impulse given balance in one direction is transmitted through lever, and whole impulse in opposite direction is transmitted directly to chronometer impulse pallet, locking and unlocking the escape-wheel but once at each impulse given by said wheel.

397. Continuous circular into intermittent rectilinear reciprocating. A motion used on several sewing machines for driving the shuttle. Same motion applied to three-revolution cylinder printing-presses.

398. Continuous circular motion into intermittent circular—the cam, C, being the driver.

399. A method of repairing chains, or tightening chains used as guys or braces. Link is made in two parts, one end of each is provided with swivel-nut, and other end with screw; the screw of each part fits into nut of other.

400. Four-motion feed (A. B. Wilson's patent), used on Wheeler & Wilson's, Sloat's, and other sewing machines. The bar, A, is forked, and has a second bar, B (carrying the spur or feeder), pivoted in the said fork. The bar, B, is lifted by a radial projection on the cam, C, at the same time the two bars are carried forward. A spring produces the return stroke, and the bar, B, drops of its own gravity.

401. E. P. Brownell's patent crank-motion to obviate dead-centers. The pressure on the treadle causes the slotted slide, A, to move forward with the wrist until the latter has passed the center, when the spring, B, forces the slide against the stops until it is again required to move forward.

402. G. O. Guernsey's patent escapement for watches. In this escapement two balance-wheels are employed, carried by the same driving-power, but oscillating in opposite directions, for the purpose of counteracting the effect of any sudden jar upon a watch or time-piece. The jar which would accelerate motion of one wheel would retard the motion of other. Anchor, A, is secured to lever, B, having an interior and exterior toothed segment at its end, each one of which gears with the pinion of balance-wheels.

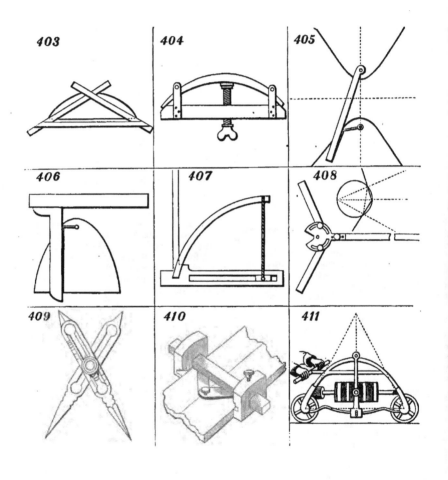

403. Cyclograph for describing circular arcs in drawings where the center is inaccessible. This is composed of three straight rules. The chord and versed sine being laid down, draw straight sloping lines from ends of former to top of latter, and to these lines lay two of the rules crossing at the apex. Fasten these rules together, and another rule across them to serve as a brace, and insert a pin or point at each end of chord to guide the apparatus, which, on being moved against these points, will describe the arc by means of pencil in the angle of the crossing edges of the sloping rules.

404. Another cyclograph. The elastic arched bar is made half the depth at the ends that it is at the middle, and is formed so that its outer edge coincides with a true circular arc when bent to its greatest extent. Three points in the required arc being given, the bar is bent to them by means of the screw, each end being confined to the straight bar by means of a small roller.

405. Mechanical means of describing hyperbolas, their foci and vertices being given. Suppose the curves two opposite hyperbolas, the points in vertical dotted center line their foci. One end of rule turns on one focus as a center through which one edge ranges. One end of thread being looped on pin inserted at the other focus, and other end held to other end of rule, with just enough slack between to permit height to reach vortex when rule coincides with center line. A pencil held in bight, and kept close to rule while latter is moved from center line, describes one-half of parabola; the rule is then reversed for the other half.

406. Mechanical means of describing parabolas, the base, altitude, focus, and directrix being given. Lay straight edge with near side coinciding with directrix, and square with stock against the same, so that the blade is parallel with the axis, and proceed with pencil in bight of thread, as in the preceding.

407. Instrument for describing pointed arches. Horizontal bar is slotted and fitted with a slide having pin for loop of cord. Arch bar of elastic wood is fixed in horizontal at right angles. Horizontal bar is placed with upper edge on springing line, and back of arch bar ranging with jamb of opening, and the latter bar is bent till the upper side meets apex of arch, fulcrum-piece at its base insuring its retaining tangential relation to jamb; the pencil is secured to arched bar at its connection with cord

408. Centrolinead for drawing lines toward an inaccessible or inconveniently distant point; chiefly used in perspective. Upper or drawing edge of blade and back of movable legs should intersect center of joint. Geometrical diagram indicates mode of setting instrument, legs forming it may form unequal angles with blade. At either end of dotted line crossing central, a pin is inserted vertically for instrument to work against. Supposing it to be inconvenient to produce the convergent lines until they intersect, even temporarily, for the purpose of setting the instrument as shown, a corresponding convergence may be found between them by drawing a line parallel to and inward from each.

409. Proportional compasses used in copying drawings on a given larger or smaller scale. The pivot of compasses is secured in a slide which is adjustable in the longitudinal slots of legs, and capable of being secured by a set screw, the dimensions are taken between one pair of points and transferred with the other pair, and thus enlarged or diminished in proportion to the relative distances of the points from the pivot. A scale is provided on one or both legs to indicate the proportion.

410. Bisecting gauge. Of two parallel cheeks on the cross-bar one is fixed and the other adjustable, and held by thumb-screw. In either cheek is centered one of two short bars of equal length, united by a pivot, having a sharp point for marking. This point is always in a central position between the cheeks, whatever their distance apart, so that any parallel sided solid to which the cheeks are adjusted may be bisected from end to end by drawing the gauge along it. Solids not parallel sided may be bisected in like manner, by leaving one cheek loose, but keeping it in contact with solid.

411. Self-recording level for surveyors. Consists of a carriage, the shape of which is governed by an isosceles triangle having horizontal base. The circumference of each wheel equals the base of the triangle. A pendulum, when the instrument is on level ground, bisects the base, and when on an inclination gravitates to right or left from center accordingly. A drum, rotated by gearing from one of the carriage wheels, carries sectionally ruled paper, upon which pencil on pendulum traces profile corresponding with amount of ground traveled over. The drum can be shifted vertically to accord with any given scale, and horizontally, to avoid removal of filled paper.

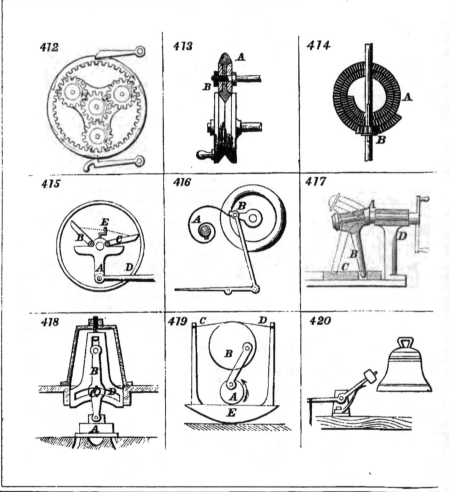

412. Wheel-work in the base of capstan. Thus provided, the capstan can be used as a simple or compound machine, single or triple purchase. The drumhead and barrel rotate independently ; the former, being fixed on spindle, turns it round, and when locked to barrel turns it also, forming single purchase ; but when unlocked, wheel-work acts, and drumhead and barrel rotate in opposite directions, with velocities as three to one.

413. J. W. Howlett's patent adjustable frictional gearing. This is an improvement on that shown in 45 of this table. The upper wheel, A, shown in section, is composed of a rubber disk with V-edge, clamped between two metal plates. By screwing up the nut, B, which holds the parts together, the rubber disk is made to expand radially, and greater tractive power may be produced between the two wheels.

414. Scroll gear and sliding pinion, to produce an increasing velocity of scroll-plate, A, in one direction, and a decreasing velocity when the motion is reversed. Pinion, B, moves on a feather on the shaft.

415. P. Dickson's patent device for converting an oscillating motion into intermittent circular, in either direction. Oscillating motion communicated to lever, A, which is provided with two pawls, B and C, hinged to its upper side, near shaft of wheel, D. Small crank, E, on upper side of lever, A, is attached by cord to each of pawls, so that when pawl, C, is let into contact with interior of rim of wheel, D, it moves in one direction, and pawl, B, is out of gear. Motion of wheel, D, may be reversed by lifting pawl, C, which was in gear, and letting opposite one into gear by crank, E.

416. A device for assisting the crank of a treadle motion over the dead-centers. The helical spring, A, has a tendency to move the crank, B, in direction at right-angles to dead-centers.

417. Continuous circular motion into a rectilinear reciprocating. The shaft, A, working in a fixed bearing, D, is bent on one end, and fitted to turn in a socket at the upper end of a rod, B, the lower end of which works in a socket in the slide, C. Dotted lines show the position of the rod, B, and slide, when the shaft has made half a revolution from the position shown in bold lines.

418. Buchanan & Righter's patent slide-valve motion. Valve, A, is attached to lower end of rod, B, and free to slide horizontally on valve-seat. Upper end of rod, B, is attached to a pin which slides in vertical slots, and a roller, C, attached to the said rod, slides in two suspended and vertically adjustable arcs, D. This arrangement is intended to prevent the valve from being pressed with too great force against its seat by the pressure of steam, and to relieve it of friction.

419. Continuous circular motion converted into a rocking motion. Used in self-rocking cradles. Wheel, A, revolves, and is connected to a wheel, B, of greater radius, which receives an oscillating motion, and wheel, B, is provided with two flexible bands, C, D, which connect each to a standard or post attached to the rocker, E, of the cradle.

420. Arrangement of hammer for striking bells. Spring below the hammer raises it out of contact with the bell after striking, and so prevents it from interfering with the vibration of the metal in the bell.

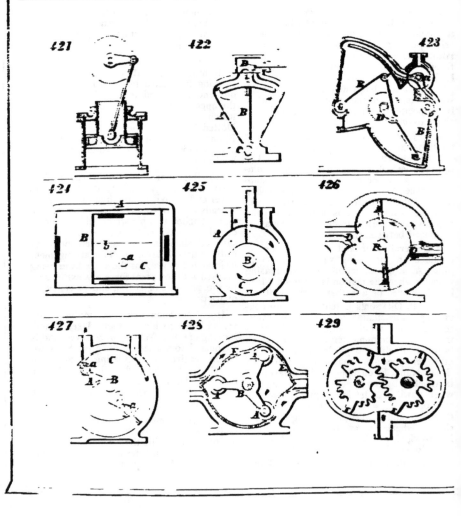

421. Trunk engine used for marine purposes. The piston has attached to it a trunk at the lower end of which the pitman is connected directly with the piston. The trunk works through a stuffing-box in cylinder-head. The effective area of the upper side of the piston is greatly reduced by the trunk. To equalize the power on both sides of piston, high-pressure steam has been first used on the upper side and afterward exhausted into and used expansively in the part of cylinder below.

422. Oscillating piston engine. The profile of the cylinder A, is of the form of a sector. The piston, B, is attached to a rock-shaft, C, and steam is admitted to the cylinder to operate on one and the other side of piston alternately, by means of a slide-valve, D, substantially like that of an ordinary reciprocating engine. The rock-shaft is connected with a crank to produce rotary motion.

423. Root's patent double-quadrant engine. This is on the same principle as 422; but two single-acting pistons, B, B, are used, and both connected with one crank, D. The steam is admitted to act on the outer sides of the two pistons alternately by means of one induction valve, a, and is exhausted through the space between the pistons. The piston and crank connections are such that the steam acts on each piston during about two-thirds of the revolution of the crank, and hence there are no dead points.

424. Root's double-reciprocating or square piston engine. The "cylinder," A, of this engine is of oblong square form and contains two pistons, B and C, the former working horizontally, and the latter working vertically within it; the piston, C, is connected with the wrist, a, of the crank on the main shaft, b. The ports for the admission of steam are shown black. The two pistons produce the rotation of the crank without dead points.

425. One of the many forms of rotary engine. A is the cylinder having the shaft, B, pass centrally through it. The piston, C, is simply an eccentric fast on the shaft and working in contact with the cylinder at one point. The induction and eduction of steam take place as indicated by arrows, and the pressure of the steam on one side of the piston produces its rotation and that of the shaft. The sliding abutment, D, between the induction and eduction ports moves out of the way of the piston to let it pass.

426. Another form of rotary engine, in which there are two stationary abutments, D, D, within the cylinder, and the two pistons, A, A, in order to enable them to pass the abutments, are made to slide radially in grooves in the hub, C, of the main shaft, B. The steam acts on both pistons at once, to produce the rotation of the hub and shaft. The induction and eduction are indicated by arrows.

427. Another rotary engine, in which the shaft, B, works in fixed bearings eccentric to the cylinder. The pistons, A, A, are fitted to slide in and out from grooves in the hub, C, which is concentric with the shaft, but they are always radial to the cylinder, being kept so by rings (shown dotted) fitting to hubs on the cylinder-heads. The pistons slide through rolling packings, a, a, in the hub, C.

428. The india-rubber rotary engine in which the cylinder has a flexible lining, E, of india-rubber, and rollers, A, A, are substituted for pistons, said rollers being attached to arms radiating from the main shaft, B. The steam acting between the india-rubber and the surrounding rigid portion of the cylinder presses the india-rubber against the rollers, and causes them to revolve around the cylinder and turn the shaft.

429. Holly's patent double-elliptical rotary engine. The two elliptical pistons geared together are operated upon by the steam entering between them, in such manner as to produce their rotary motion in opposite directions.

These rotary engines can all be converted into pumps.

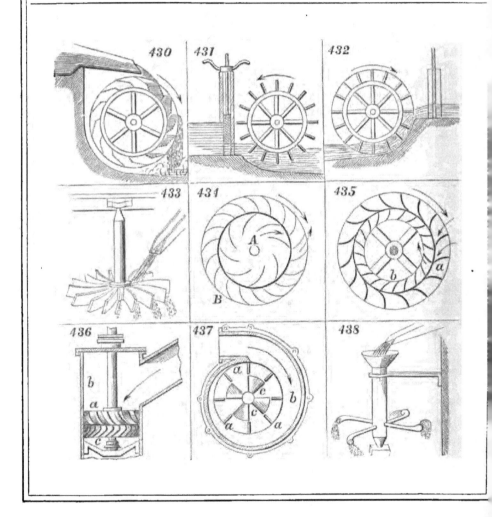

430. Overshot water-wheel.

431. Undershot water-wheel.

432. Breast-wheel. This holds intermediate place between overshot and undershot wheels; has float-boards like the former, but the cavities between are converted into buckets by moving in a channel adapted to circumference and width, and into which water enters nearly at the level of axle.

433. Horizontal overshot water-wheel.

434. A plan view of the Fourneyron turbine water-wheel. In the center are a number of fixed curved "shutes" or guides, A, which direct the water against the buckets of the outer wheel, B, which revolves, and the water discharges at the circumference.

435. Warren's central discharge turbine, plan view. The guides, a, are outside, and the wheel, b, revolves within them, discharging the water at the center.

436. Jonval turbine. The "shutes" are arranged on the outside of a drum, radial to a common center and stationary within the trunk or casing, b. The wheel, c, is made in nearly the same way; the buckets exceed in number those of the shutes, and are set at a slight tangent instead of radially, and the curve generally used is that of the cycloid or parabola.

437. Volute wheel, having radial vanes, a, against which the water impinges and carries the wheel around. The scroll or volute casing, b, confines the water in such a manner that it acts against the vanes all around the wheel. By the addition of the inclined buckets, c, c, at the bottom, the water is made to act with additional force as it escapes through the openings of said buckets.

438. Barker's or reaction mill. Rotary motion of central hollow shaft is obtained by the reaction of the water escaping at the ends of its arms, the rotation being in a direction the reverse of the escape.

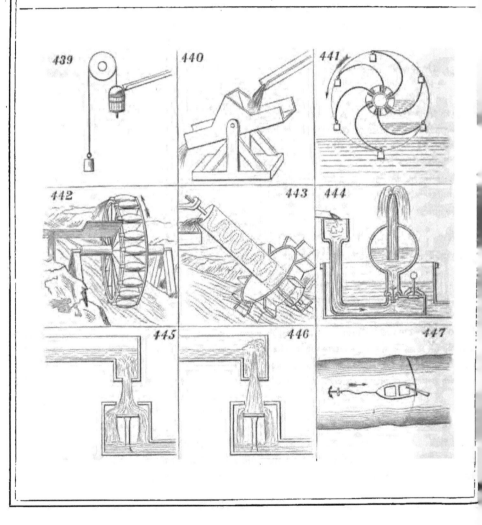

439. A method of obtaining a reciprocating motion from a continuous fall of water, by means of a valve in the bottom of the bucket which opens by striking the ground and thereby emptying the bucket, which is caused to rise again by the action of a counter-weight on the other side of the pulley over which it is suspended.

440. Represents a trough divided transversely into equal parts and supported on an axis by a frame beneath. The fall of water filling one side of the division, the trough is vibrated on its axis, and at the same time that it delivers the water the opposite side is brought under the stream and filled, which in like manner produces the vibration of the trough back again. This has been used as a water meter.

441. Persian wheel, used in Eastern countries for irrigation. It has a hollow shaft and curved floats, at the extremities of which are suspended buckets or tubs. The wheel is, partly immersed in a stream acting on the convex surface of its floats, and as it is thus caused to revolve, a quantity of water will be elevated by each float at each revolution, and conducted to the hollow shaft at the same time that one of the buckets carries its fill of water to a higher level, where it is emptied by coming in contact with a stationary pin placed in a convenient position for tilting it.

442. Machine of ancient origin, still employed on the river Eisach, in the Tyrol, for raising water. A current keeping the wheel in motion, the pots on its periphery are successively immersed, filled, and emptied into a trough above the stream.

443. Application of Archimedes's screw to raising water, the supply stream being the motive power. The oblique shaft of the wheel has extending through it a spiral passage, the lower end of which is immersed in water, and the stream, acting upon the wheel at its lower end, produces its revolution, by which the water is conveyed upward continuously through the spiral passage and discharged at the top.

444. Montgolfier's hydraulic ram. Small fall of water made to throw a jet to a great height or furnish a supply at high level. The right-hand valve being kept open by a weight or spring, the current flowing through the pipe in the direction of the arrow escapes thereby till its pressure, overcoming the resistance of weight or spring, closes it. On the closing of this valve the momentum of the current overcomes the pressure on the other valve, opens it, and throws a quantity of water into the globular air-chamber by the expansive force of the air in which the upward stream from the nozzle is maintained. On equilibrium taking place, the right-hand valve opens and left-hand one shuts. Thus, by the alternate action of the valves, a quantity of water is raised into the air-chamber at every stroke, and the elasticity of the air gives uniformity to the efflux.

445 and 446. D'Ectol's oscillating column, for elevating a portion of a given fall of water above the level of the reservoir or head, by means of a machine all the parts of which are absolutely fixed. It consists of an upper and smaller tube, which is constantly supplied with water, and a lower and larger tube, provided with a circular plate below concentric with the orifice which receives the stream from the tube above. Upon allowing the water to descend as shown in 445, it forms itself gradually into a cone on the circular plate, as shown in 446, which cone protrudes into the smaller tube so as to check the flow of water downward; and the regular supply continuing from above, the column in the upper tube rises until the cone on the circular plate gives way. This action is renewed periodically and is regulated by the supply of water.

447. This method of passing a boat from one shore of a river to the other is common on the Rhine and elsewhere, and is effected by the action of the stream on the rudder, which carries the boat across the stream in the arc of a circle, the center of which is the anchor which holds the boat from floating down the stream.

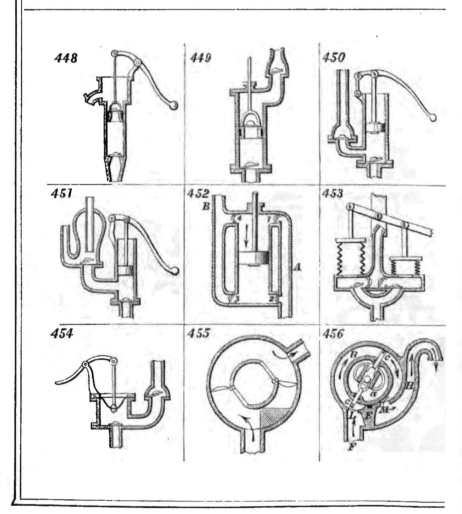

448. Common lift pump. In the up-stroke of piston or bucket the lower valve opens and the valve in piston shuts ; air is exhausted out of suction-pipe, and water rushes up to fill the vacuum. In down-stroke, lower valve is shut and valve in piston opens, and the water simply passes through the piston. The water above piston is lifted up, and runs over out of spout at each up-stroke. This pump cannot raise water over thirty feet high.

449. Modern lifting pump. This pump operates in same manner as one in previous figure, except that piston-rod passes through stuffing-box, and outlet is closed by a flap-valve opening upward. Water can be lifted to any height above this pump.

450. Ordinary force pump, with two valves. The cylinder is above water, and is fitted with solid piston ; one valve closes outlet-pipe, and other closes suction-pipe. When piston is rising suction-valve is open, and water rushes into cylinder, outlet-valve being closed. On descent of piston suction-valve closes, and water is forced up through outlet-valve to any distance or elevation.

451. Force pump, same as above, with addition of air-chamber to the outlet, to produce a constant flow. The outlet from air-chamber is shown at two places, from either of which water may be taken. The air is compressed by the water during the downward stroke of the piston, and expands and presses out the water from the chamber during the up-stroke.

452. Double-acting pump. Cylinder closed at each end, and piston-rod passes through stuffing-box on one end, and the cylinder has four openings covered by valves, two for admitting water and like number for discharge. A is suction-pipe, and B discharge-pipe. When piston moves down, water rushes in at suction-valve, 1, on upper end of cylinder, and that below piston is forced through valve, 3, and discharge-pipe, B ; on the piston ascending again, water is forced through discharge-valve, 4, on upper end of cylinder, and water enters lower suction-valve, 2.

453. Double lantern-bellows pump. As one bellows is distended by lever, air is rarefied within it, and water passes up suction-pipe to fill space ; at same time other bellows is compressed, and expels its contents through discharge-pipe ; valves working the same as in the ordinary force pump.

454. Diaphragm forcing pump. A flexible diaphragm is employed instead of bellows, and valves are arranged same as in preceding.

455. Old rotary pump. Lower aperture entrance for water, and upper for exit. Central part revolves with its valves, which fit accurately to inner surface of outer cylinder. The projection shown in lower side of cylinder is an abutment to close the valves when they reach that point.

456. Cary's rotary pump. Within the fixed cylinder there is placed a revolving drum, B, attached to an axle, A. Heart-shaped cam, a, surrounding axle, is also fixed. Revolution of drum causes sliding-pistons, c, c, to move in and out in obedience to form of cam. Water enters and is removed from the chamber through ports, L and M ; the directions are indicated by arrows. Cam is so placed that each piston is, in succession, forced back to its seat when opposite E, and at same time other piston is forced fully against inner side of chamber, thus driving before it water already there into exit-pipe, H, and drawing after it through suction-pipe, F, the stream of supply.

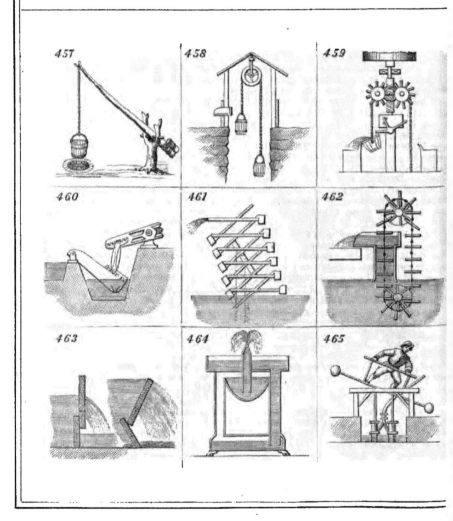

457. Common mode of raising water from wells of inconsiderable depth. Counter-balance equals about one-half of weight to be raised, so that the bucket has to be pulled down when empty, and is assisted in elevating it when full by counterbalance.

458. The common pulley and buckets for raising water; the empty bucket is pulled down to raise the full one.

459. Reciprocating lift for wells. Top part represents horizontal wind-wheel on a shaft which carries spiral thread. Coupling of latter allows small vibration, that it may act on one worm-wheel at a time. Behind worm-wheels are pulleys over which passes rope which carries bucket at each extremity. In center is vibrating tappet, against which bucket strikes in its ascent, and which, by means of arm in step wherein spiral and shaft are supported, traverses spiral from one wheel to other so that the bucket which has delivered water is lowered and other one raised.

460. Fairbairn's bailing-scoop, for elevating water short distances. The scoop is connected by pitman to end of a lever or of a beam of single-acting engine Distance of lift may be altered by placing end of rod in notches shown in figure.

461. Pendulums or swinging gutters for raising water by their pendulous motions. Terminations at bottom are scoops, and at top open pipes; intermediate angles are formed with boxes (and flap valve), each connected with two branches of pipe.

462. Chain pump; lifting water by continuous circular motion. Wood or metal disks, carried by endless chain, are adapted to water-tight cylinder, and form with it a succession of buckets filled with water. Power is applied at upper wheel.

463. Self-acting weir and scouring sluice. Two leaves turn on pivots below centers; upper leaf much larger than lower, and turns in direction of stream, while lower turns against it. Top edge of lower leaf overlaps bottom edge of upper one and is forced against it by pressure of water. In ordinary states of stream, counteracting pressures keep weir vertical and closed, as in the left-hand figure, and water flows through notch in upper leaf; but on water rising above ordinary level, pressure above from greater surface and leverage overcomes resistance below, upper leaf turns over, pushing back lower, reducing obstructions and opening at bed a passage to deposit.

464. Hiero's fountain. Water being poured into upper vessel descends tube on right into lower; intermediate vessel being also filled and more water poured into upper, confined air in cavities over water in lower and intermediate vessels and in communication tube on left, being compressed, drives by its elastic force a jet up central tube.

465. Balance pumps. Pair worked reciprocally by a person pressing alternately on opposite ends of lever or beam.

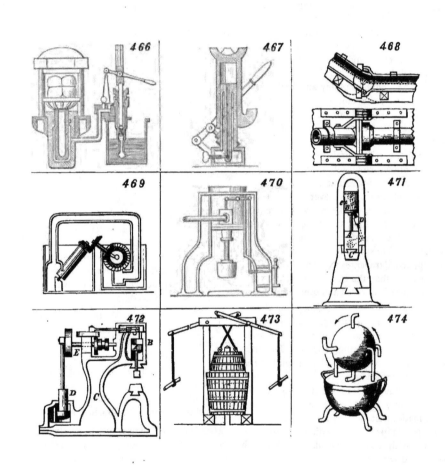

466. Hydrostatic press. Water forced by the pump through the small pipe into the ram cylinder and under the solid ram, presses up the ram. The amount of force obtained is in proportion to the relative areas or squares of diameters of the pump-plunger and ram. Suppose, for instance, the pump-plunger to be one inch diameter and the ram thirty inches, the upward pressure received by the ram would be 900 times the downward pressure of the plunger.

467. Robertson's hydrostatic jack. In this the ram is stationary upon a hollow base and the cylinder with claw attached slides upon it. The pump takes the water from the hollow base and forces it through a pipe in the ram into the cylinder, and so raises the latter. At the bottom of pipe there is a valve operated by a thumb-screw to let back the water and lower the load as gradually as may be desired.

468. Flexible water main, plan and section. Two pipes of 15 and 18 inches interior diameter, having some of their joints thus formed, conduct water across the Clyde to Glasgow Water-works. Pipes are secured to strong log frames, having hinges with horizontal pivots. Frames and pipes were put together on south side of the river, and, the north end of pipe being plugged, they were hauled across by machinery on north side, their flexible structure enabling them to follow the bed.

469. French invention for obtaining rotary motion from different temperatures in two bodies of water. Two cisterns contain water : that in left at natural temperature and that in right higher. In right is a water-wheel geared with Archimedean screw in left. From spiral screw of the latter a pipe extends over and passes to the under side of wheel. Machine is started by turning screw in opposite direction to that for raising water, thus forcing down air, which ascends in tube, crosses and descends, and imparts motion to wheel ; and its volume increasing with change of temperature, it is said, keeps the machine in motion. We are not informed how the difference of temperature is to be maintained.

470. Steam hammer. Cylinder fixed above and hammer attached to lower end of piston-rod.

Steam being alternately admitted below piston and allowed to escape, raises and lets fall the hammer.

471. Hotchkiss's atmospheric hammer ; derives the force of its blow from compressed air. Hammer head, C, is attached to a piston fitted to a cylinder, B, which is connected by a rod, D, with a crank, A, on the rotary driving-shaft. As the cylinder ascends, air entering hole, e, is compressed below piston and lifts hammer. As cylinder descends, air entering hole, e, is compressed above and is stored up to produce the blow by its instant expansion after the crank and connecting-rod turn bottom center.

472. Grimshaw's compressed air hammer. The head of this hammer is attached to a piston, A, which works in a cylinder, B, into which air is admitted—like steam to a steam engine— above and below the piston by a slide-valve on top. The air is received from a reservoir, C, in the framing, supplied by an air pump, D, driven by a crank on the rotary driving-shaft, E.

473. Air-pump of simple construction. Smaller tub inverted in larger one. The latter contains water to upper dotted line, and the pipe from shaft or space to be exhausted passes through it to a few inches above water, terminating with valve opening upward. Upper tub has short pipe and upwardly-opening valve at top, and is suspended by ropes from levers. When upper tub descends, great part of air within is expelled through upper valve, so that, when afterward raised, rarefaction within causes gas or air to ascend through the lower valve. This pump was successfully used for drawing off carbonic acid from a large and deep shaft.

474. Æolipile or Hero's steam toy, described by Hero, of Alexandria, 130 years B.C., and now regarded as the first steam engine, the rotary form of which it may be considered to represent. From the lower vessel, or boiler, rise two pipes conducting steam to globular vessel above, and forming pivots on which the said vessel is caused to revolve in the direction of arrows, by the escape of steam through a number of bent arms. This works on the same principle as Barker's mill, 438 in this table.

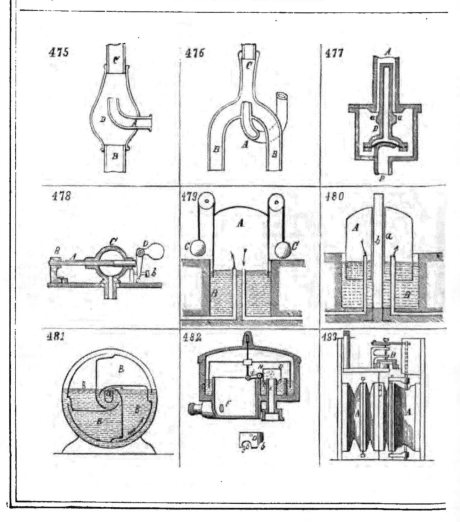

475. Bilge ejector (Brear's patent) for discharging bilge-water from vessels, or for raising and forcing water under various circumstances. D is a chamber having attached a suction-pipe, B, and discharge-pipe, C, and having a steam-pipe entering at one side, with a nozzle directed toward the discharge-pipe. A jet of steam entering through A expels the air from D and C, produces a vacuum in B, and causes water to rise through B, and pass through D and C, in a regular and constant stream. Compressed air may be used as a substitute for steam.

476. Another apparatus operating on the same principle as the foregoing. It is termed a steam siphon pump (Lansdell's patent). A is the jet-pipe ; B, B, are two suc-tion-pipes, having a forked connection with the discharge-pipe, C. The steam jet-pipe entering at the fork offers no obstacle to the upward passage of the water, which moves upward in an unbroken current.

477. Steam trap for shutting in steam, but providing for the escape of water from steam coils and radiators (Hoard & Wiggin's patent). It consists of a box, connected at A with the end of the coil or the waste-pipe, having an outlet at B, and furnished with a hollow valve, D, the bottom of which is composed of a flexible diaphragm. Valve is filled with liquid, and hermetically sealed, and its diaphragm rests upon a bridge over the outlet-pipe. The presence of steam in the outer-box so heats the water in valve that the diaphragm expands and raises valve up to the seat, a, a. Water of condensation accumulating reduces the tempera-ture of valve ; and as the liquid in valve contracts, dia-phragm allows valve to descend and let water off.

478. Another steam trap (Ray's patent). Valve, a, closes and opens by longitudinal expansion and contraction of waste-pipe, A, which terminates in the middle of an at-tached hollow sphere, C. A portion of the pipe is firmly secured to a fixed support, B. Valve consists of a plunger which works in a stuffing-box in the sphere, opposite the end of the pipe, and it is pressed toward the end of the pipe by a loaded elbow lever, D, as far as permitted by a stop-screw, b, and stop, c. When pipe is filled with water, its length is so reduced that valve remains open ; but when filled with steam, it is expanded so that valve closes it. Screw, b, serves to adjust the action of valve.

479. Gasometer. The open-bottomed vessel, A, is ar-ranged in the tank, B, of water, and partly counterbalanced by weights, C, C. Gas enters the gasometer by one and leaves it by the other of the two pipes inserted through the bottom of the tank. As gas enters, vessel, A, rises, and vice versa. The pressure is regulated by adding to or reducing the weights, C, C.

480. Another kind of gasometer. The vessel, A, has permanently secured within it a central tube, a, which slides on a fixed tube, b, in the center of the tank.

481. Wet gas meter. The stationary case, A, is filled with water up to above the center. The inner revolving drum is divided into four compartments, B, B, with inlets around the central pipe, a, which introduces the gas through one of the hollow journals of the drum. This pipe is turned up to admit the gas above the water, as indi-cated by the arrow near the center of the figure. As gas enters the compartments, B, B, one after another, it turns the drum in the direction of the arrow shown near its peri-phery, displacing the water from them. As the chambers pass over they fill with water again. The cubic contents of the compartments being known, and the number of revolutions of the drum being registered by dial-work, the quantity of gas passing through the meter is registered.

482. Gas regulator (Powers's patent) for equalizing the supply of gas to all the burners of a building or apartment, notwithstanding variations in the pressure on the main, or variations produced by turning gas on or off, to or from any number of the burners. The regulator-valve, D, of which a separate outside view is given, is arranged over inlet-pipe, E, and connected by a lever, d, with an inverted cup, H, the lower edges of which, as well as those of valve, dip into channels containing quicksilver. There is no escape of gas around the cup, H, but there are notches, b, in the valve to permit the gas to pass over the surface of the quicksilver. As the pressure of gas increases, it acts upon the inner surface of cup, H, which is larger than valve, and the cup is thereby raised, causing a depression of the valve into the quicksilver, and contracting the opening notches, b, and diminishing the quantity of gas passing through. As the pressure diminishes, an opposite result is produced. The outlet to burners is at F.

483. Dry gas meter. Consists of two bellows-like cham-bers, A, A′, which are alternately filled with gas, and dis-charged through a valve, B, something like the slide-valve of a steam engine, worked by the chambers, A, A′. The capacity of the chambers being known, and the number of times they are filled being registered by dial-work, the quantity of gas passing through the meter is indicated on the dials.

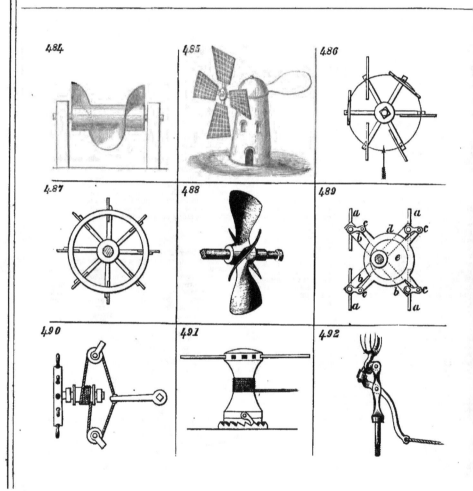

484. A spiral wound round a cylinder to convert the motion of the wind or a stream of water into rotary motion.

485. Common wind-mill, illustrating the production of circular motion by the direct action of the wind upon the oblique sails.

486. Plan of a vertical wind-mill. The sails are so pivoted as to present their edges in returning toward the wind, but to present their faces to the action of the wind, the direction of which is supposed to be as indicated by the arrow.

487. Common paddle-wheel for propelling vessels; the revolution of the wheel causes the buckets to press backward against the water and so produce the forward movement of the vessel.

488. Screw propeller. The blades are sections of a screw-thread, and their revolution in the water has the same effect as the working of a screw in a nut, producing motion in the direction of the axis and so propelling the vessel.

489. Vertical bucket paddle-wheel. The buckets, a, a, are pivoted into the arms, b, b, at equal distances from the shaft. To the pivots are attached cranks, c, c, which are pivoted at their ends to the arms of a ring, d, which is fitted loosely to a stationary eccentric, e. The revolution of the arms and buckets with the shaft causes the ring, d, also to rotate upon the eccentric, and the action of this ring on the cranks keeps the buckets always upright, so that they enter the water and leave it edgewise without re-

sistance or lift, and while in the water are in the most effective position for propulsion.

490. Ordinary steering apparatus. Plan view. On the shaft of the hand-wheel there is a barrel on which is wound a rope which passes round the guide-pulleys and has its opposite ends attached to the "tiller" or lever on the top of the rudder; by turning the wheel, one end of the rope is wound on and the other let off, and the tiller is moved in one or the other direction, according to the direction in which the wheel is turned.

491. Capstan. The cable or rope wound on the barrel of the capstan is hauled in by turning the capstan on its axis by means of hand-spikes or bars inserted into holes in the head. The capstan is prevented from turning back by a pawl attached to its lower part and working in a circular ratchet on the base.

492. Boat-detaching hook (Brown & Level's). The upright standard is secured to the boat, and the tongue hinged to its upper end enters an eye in the level which works on a fulcrum at the middle of the standard. A similar apparatus is applied at each end of the boat. The hooks of the tackles hook into the tongues, which are secure until it is desired to detach the boat, when a rope attached to the lower end of each lever is pulled in such a direction as to slip the eye at the upper end of the lever from off the tongue, which being then liberated slips out of the hook of the tackle and detaches the boat.

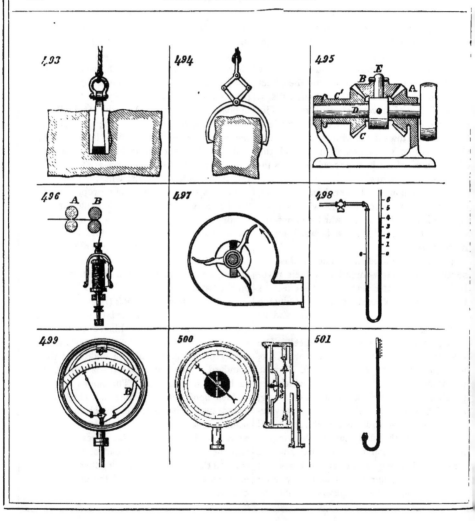

493. "Lewis." for lifting stone in building. It is composed of a central taper pin or wedge, with two wedge-like packing-pieces arranged one on each side of it. The three pieces are inserted together in a hole drilled into the stone, and when the central wedge is hoisted upon it wedges the packing-pieces out so tightly against the sides of the hole as to enable the stone to be lifted.

494. Tongs for lifting stones, etc. The pull on the shackle which connects the two links causes the latter so to act on the upper arms of the tongs as to make their points press themselves against or into the stone. The greater the weight the harder the tongs bite.

- 495. Entwistle's patent gearing. Bevel-gear, A, is fixed. B, gearing with A, is fitted to rotate on stud, E, secured to shaft, D, and it also gears with bevel-gear, C, loose, on the shaft, D. On rotary motion being given to shaft, D, the gear, E, revolves around A, and also rotates upon its own axis, and so acts upon C in two ways, namely, by its rotation on its own axis and by its revolution around A. With three gears of equal size, the gear, C, makes two revolutions for every one of the shaft, D. This velocity of revolution may, however, be varied by changing the relative sizes of the gears. C is represented with an attached drum, C'. This gearing may be used for steering apparatus, driving screw-propellers, etc. By applying power to C, action may be reversed, and a slow motion of D obtained.

496. Drawing and twisting in spinning cotton, wool, etc. The front drawing-rolls, B, rotate faster than the back ones, A, and so produce a draught, and draw out the fibers of the sliver or roving passing between them. Roving passes from the front drawing-rolls to throstle, which, by its rotation around the bobbin, twists and winds the yarn on the bobbin.

497. Fan-blower. The casing has circular openings in its sides through which, by the revolution of the shaft and attached fan-blades, air is drawn in at the center of the casing, to be forced out under pressure through the spout.

498. Siphon pressure gauge. Lower part of bent tube contains mercury. The leg of the tube, against which the scale is marked, is open at top, the other leg connected with the steam-boiler or other apparatus on which the pressure is to be indicated. The pressure on the mercury in the one leg causes it to be depressed in that and raised in the other until there is an equilibrium established between the weight of mercury and pressure of steam in one leg, and the weight of mercury and pressure of atmosphere in the other. This is the most accurate gauge known; but as high pressure requires so long a tube, it has given place to those which are practically accurate enough, and of more convenient form.

499. Aneroid gauge, known as the "Bourdon gauge," from the name of its inventor, a Frenchman. B is a bent tube closed at its ends, secured at C, the middle of its length, and having its ends free. Pressure of steam or other fluid admitted to tube tends to straighten it more or less, according to its intensity. The ends of tube are connected with a toothed sector-piece gearing, with a pinion on the spindle of a pointer which indicates the pressure on a dial.

500. Pressure gauge now most commonly used. Sometimes known as the "Magdeburg gauge," from the name of the place where first manufactured. Face view and section. The fluid whose pressure is to be measured acts upon a circular metal disk, A, generally corrugated, and the deflection of the disk under the pressure gives motion to a toothed sector, e, which gears with a pinion on the spindle of the pointer.

501. Mercurial barometer. Longer leg of bent tube, against which is marked the scale of inches, is closed at top, and shorter one is open to the atmosphere, or merely covered with some porous material. Column of mercury in longer leg, from which the air has been extracted, is held up by the pressure of air on the surface of that in the shorter leg, and rises or falls as the pressure of the atmosphere varies. The old-fashioned weather-glass is composed of a similar tube attached to the back of a dial, and a float inserted into the shorter leg of the tube, and geared by a rack and pinion, or cord and pulley, with the spindle of the pointer.

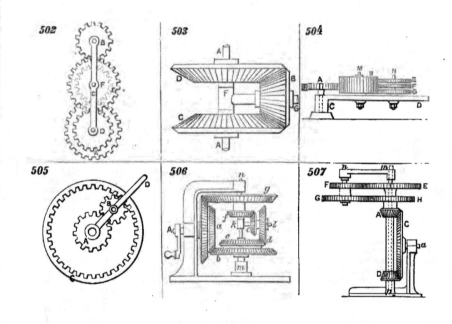

502. An "epicyclic train." Any train of gearing the axes of the wheels of which revolve around a common center is properly known by this name. The wheel at one end of such a train, if not those at both ends, is always concentric with the revolving frame. C is the frame or train-bearing arm. The center wheel, A, concentric with this frame, gears with a pinion, F, to the same axle with which is secured a wheel, E, that gears with a wheel, B. If the first wheel, A, be fixed and a motion be given to the frame, C, the train will revolve around the fixed wheel and the relative motion of the frame to the fixed wheel will communicate through the train a rotary motion to B on its axis. Or the first wheel as well as the frame may be made to revolve with different velocities, with the same result except as to the velocity of rotation of B upon its axis.

In the epicyclic train as thus described only the wheel at one extremity is concentric with the revolving frame; but if the wheel, E, instead of gearing with B, be made to gear with the wheel, D, which like the wheel, A, is concentric with the frame, we have an epicyclic train of which the wheels at both extremities are concentric with the frame. In this train we may either communicate the driving motion to the arm and one extreme wheel, in order to produce an aggregate rotation of the other extreme wheel, or motion may be given to the two extreme wheels, A and D, of the train, and the aggregate motion will thus be communicated to the arm.

503. A very simple form of the epicyclic train, in which F, G, is the arm, secured to the central shaft, A, upon which are loosely fitted the bevel-wheels, C, D. The arm is formed into an axle for the bevel-wheel, B, which is fitted to turn freely upon it. Motion may be given to the two wheels, C, D, in order to produce aggregate motion of the arm, or else to the arm and one of said wheels in order to produce aggregate motion of the other wheel.

504. "Ferguson's mechanical paradox," designed to show a curious property of the epicyclic train. The wheel, A, is fixed upon a stationary stud about which the arm, C, D, revolves. In this arm are two pins, M, N, upon one of which is fitted loosely a thick wheel, B, gearing with A, and upon the other are three loose wheels, E, F, G, all gearing with B. When the arm, C, D, is turned round on the stud, motion is given to the three wheels, E, F, G, on their common axis, viz., the pin, N; the three forming with the intermediate wheel, B, and the wheel, A, three distinct epicyclic trains. Suppose A to have twenty teeth, F twenty, E twenty-one, and G nineteen; as the arm, E, C, D, is turned round, F will appear not to turn on its axis, as any point in its circumference will always point in one direction, while E will appear to turn slowly in one and G in the other direction, which—an apparent paradox—gave rise to the name of the apparatus.

505. Another simple form of the epicyclic train, in which the arm, D, carries a pinion, B, which gears both with a spur-wheel, A, and an annular wheel, C, both concentric with the axis of the arm. Either of the wheels, A, C, may be stationary, and the revolution of the arm and pinion will give motion to the other wheel.

506. Another epicyclic train in which neither the first nor last wheel is fixed. $m, n,$ is a shaft to which is firmly secured the train-bearing arm, $k, l,$ which carries the two wheels, $d, e,$ secured together, but rotating upon the arm itself. The wheels, b and $c,$ are united and turn together, freely upon the shaft, $m, n;$ the wheels, f and $g,$ are also secured together, but turn together freely on the shaft, $m, n.$ The wheels, c, d, e and $f,$ constitute an epicyclic train of which c is the first and f the last wheel. A shaft, A, is employed as a driver, and has firmly secured to it two wheels, a and $h,$ the first of which gears with the wheel, $b,$ and thus communicates motion to the first wheel, $c,$ of the epicyclic train, and the wheel, $h,$ drives the wheel, $g,$ which thus gives motion to the last wheel, $f.$ Motion communicated in this way to the two ends of the train produces an aggregate motion of the arm, $k, l,$ and shaft, $m, n.$

This train may be modified; for instance, suppose the wheels, g and $f,$ to be disunited, g to be fixed to the shaft, $m, n,$ and f only running loose upon it. The driving-shaft, A, will as before communicate motion to the first wheel, $c,$ of the epicyclic train by means of the wheels, a and $b,$ and will also by h cause the wheel, $g,$ the shaft, $m, n,$ and the train-bearing arm, $k, l,$ to revolve, and the aggregate rotation will be given to the loose wheel, $f.$

507. Another form of epicyclic train designed for producing a very slow motion. m is a fixed shaft upon which is loosely fitted a long sleeve, to the lower end of which is fixed a wheel, D, and to the upper end a wheel, E. Upon this long sleeve there is fitted a shorter one which carries at its extremities the wheels, A and H. A wheel, C, gears with both D and A, and a train-bearing arm, $m, n,$ which revolves freely upon the shaft, $m, p,$ carries upon a stud at n the united wheels, F and G. If A have 10 teeth, C 100, D 10, E 61, F, 49, G 41, and H 51, there will be 25,000 revolutions of the train-bearing arm, $m, n,$ for one of the wheel, C.

COLLINS & CO'S
NEW PATENT SCREW-WRENCH.

ALL MECHANICS,

Will have noticed that their wrenches that fail by the bearing back of the handle, the failure of which is produced in our invention.

ADVANTAGES.

1st. The screw stock, A, in which the screw works, is a separate piece from the other part, and is made independent on the wrench bar, C, by the use of a thimble, D, which also prevents all strain on the small nut, B, at the end of the handle.

2d. The wrench bar, C, is made much stronger than those of other wrenches in market, preventing their springing and thereby impairing their working.

3d. ALL OF THE PARTS ARE CASE HARDENED, adding to their stiffness, and preventing their being easily bruised.

4th. They are made by new and improved machinery, in such a manner that all the parts are INTERCHANGEABLE. This is a new feature, and will be highly appreciated by machinists and others being many wrenches, for it one part should fail or wear out, that piece can be supplied without being obliged to purchase a new wrench.

Solid Cast Steel Wrenches, of above patent, made to order.

For sale by Hardware Dealers throughout the United States.   Address,

COLLINS & CO.

EXTENSIONS OF PATENTS.

INVENTORS of new and useful discoveries or improvements patented in 1854, or at any subsequent period prior to March 2, 1861, or their heirs, will do well to observe that by the law of the United States, patents granted within the periods here stated may, upon proper grounds, be extended for a period of seven years in favor of *the inventor or his heirs;* the prohibition as to extension only applying to patents issued, since March 2, 1861. The novelty, at the time of the patent having been granted, and the utility and value and importance of the invention to the public being proved, to obtain an extension for a period of seven years is only necessary for the inventor (or his heirs) to show that, having used diligence and without fault on his (or their) part, an adequate remuneration has not been received. The mere fact of an inventor having sold his patent or parted with his right for an adequate amount, so that he has been excluded from making the invention profitable to himself, however well it may have paid others, does not debar him from obtaining an extension. In 1854 nearly 2,000 patents were issued, say 1,800 for new and useful mechanical inventions, any or all of which are, or were, open to extension, provided reasonable grounds for the same, as already mentioned, can or could be shown. With many of these, however, it is now too late, as all of said patents expire during the current year; and the law requires that the petition for extension should be filed at least ninety days before the expiration of the patent; such cases should, in fact, be put in competent hands to prosecute some four to six months in advance thereof. Many of these patents are extremely valuable on account of the original ground they cover, occasionally making more recent patents subsidiary to them; and holders of patents issued in 1854–55 ought now to be making their necessary arrangements for extension. BROWN, COOMBS & Co., Solicitors of Patents, 189 Broadway, New York, are always ready to give advice in such applications and spare no pains in prosecuting them.

FOREIGN PATENTS.

THE patent systems of the various countries of Europe differ in so many essential respects not only from that of the United States, but from each other, that although the applications for patents in those countries must be conducted by attorneys or solicitors in their respective capitals, it is very necessary for all Americans intending to apply for European patents to first secure the advice and aid of thoroughly competent agents in this country, so that their applications may be put in proper condition for transmission.

For foreign patents models are not required, and hence the total cost of the patents is limited to the Government and agency fees. There are many cases in which, in European countries, as much matter relating to one subject can be embraced in a single patent as would require two or more separate and distinct patents in the United States. It is important that this should be borne in mind by American inventors, and that in order to secure the greatest possible protection at the least expense, they should ask the advice only of such solicitors as are perfectly familiar with foreign practice, and capable of judging how much would be allowed under one patent. In many cases patents in England and France, and some of the other more important countries in Europe, can be secured at a cost in each country less than would, with cost of models, be involved in patenting the same amount of subject-matter in the United States.

MESSRS. BROWN, COOMBS & Co., of the "AMERICAN ARTISAN PATENT AGENCY," are very extensively engaged in procuring foreign patents. They have at their command such experience and business facilities as will enable them to prepare in the best possible manner, and as expeditiously as is practicable, such drawings, specifications, and other documents as are required. It may be here stated that the senior member of their firm, Mr. HENRY T. BROWN, during more than *twenty-two years'* practice, has prepared more foreign applications than any practitioner in the United States.

MESSRS. BROWN, COOMBS & Co. have also so organized their relations with the most experienced and reliable patent attorneys in the principal capitals of Europe, who are now acting as their agents, that they are enabled to insure the best attention to the application of their clients while in progress, and to meet at once any obstacles or difficulties that may arise.

A circular stating the cost of patents in the principal foreign countries, and containing other valuable information on the subject, may be obtained by addressing—

BROWN, COOMBS & CO., Solicitors of Patents,
No. 189 BROADWAY, NEW YORK.

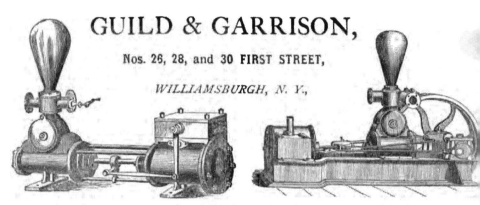

B. T. BABBITT'S LION COFFEE.

OFFICE OF B. T. BABBITT,

Nos. 64, 65, 66, 67, 68, 69, 70, 72, *and* 74 *Washington, Street,*

NEW YORK.

I am putting into the Market a superior article of PURE JAVA COFFEE, HERMETI-
LLY ROASTED, GROUND, AND SEALED *in One Pound Cans.*

I will give ONE OUNCE OF GOLD for every ounce of adulteration found in my Coffee.

For the first six months I shall put THREE ONE-DOLLAR GREENBACKS *in every*
x *of Sixty Pounds, (that is, three cans in each box will contain* A DOLLAR BILL.)

In the usual way of roasting, you can smell the coffee a long distance from the mill while
asting, which proves that there is a *very large* percentage of the *aroma* or *flavor lost,* which
the richest and best part of the coffee.

THIS AROMA OR FLAVOR I SAVE.

I have patented a new Roasting Machine, whereby the Coffee is HERMETICALLY (without
ange of air) ROASTED, GROUND, and SEALED in cans; consequently, EVERY PARTICLE of
e AROMA is SAVED. My manner of roasting gives the Coffee a rich, glossy appearance,
oduced by CONDENSING the AROMA. This brings the Coffee to the highest state of per-
tion.

I can, with confidence, recommend this Coffee as being STRICTLY PURE, and from fifteen
twenty per cent STRONGER IN AROMA than any other pure Java Coffee. It needs only
e trial to become permanently used in every family that appreciate a GOOD CUP OF COFFEE.

Yours respectfully,

B. T. BABBITT.

CIRCULAR SAWS,

WITH EMERSON'S PATENT MOVABLE TEETH

These Saws are meeting with UNPRECEDENTED SUCCESS. Their GREAT SUPERIORITY OVER EVERY OTHER KIND, both as to EFFICIENCY AND ECONOMY, is now fully established. Also,

Emerson's Patent Perforated Circular and Long Saws.

(All Gumming avoided.)

EMERSON'S PATENT ADJUSTABLE SWAGE

For Spreading, Sharpening, and Shaping the Teeth of all Splitting Saws. Price, $5.

Manufactured by the

AMERICAN SAW COMPANY

Office, No. 2 Jacob Street, near Ferry Street, New York.

Factory, Trenton, N. J.

☞ Send for new Descriptive Pamphlet and Price List.

CPSIA information can be obtained
at www.ICGtesting.com
Printed in the USA
BVHW040216271120
594322BV00007B/47